CULTURE, COMMUNICATION, AND CYBERSPACE

Rethinking Technical Communication for International Online Environments

Edited by

Kirk St.Amant
East Carolina University

and

Filipp Sapienza
White Mouse Solutions, LLC.

Baywood's Technical Communications Series
Series Editor: CHARLES H. SIDES

Routledge
Taylor & Francis Group
LONDON AND NEW YORK

First published 2011 by Baywood Publishing Company, Inc.

2 Park Square, Milton Park, Abingdon, Oxon OX14 4RN
711 Third Avenue, New York, NY 10017, USA

Routledge is an imprint of the Taylor & Francis Group, an informa business

First issued in paperback 2017

Library of Congress Catalog Number: 2010001620
ISBN 13: 978-0-89503-398-7 (hbk)

Library of Congress Cataloging-in-Publication Data

Culture, communication, and cyberspace : rethinking technical communication for international online environments / edited by Kirk St. Amant and Filipp Sapienza.
p. cm. -- (Baywood's technical communications series)
Includes bibliographical references and index.
ISBN 978-0-89503-398-7 (cloth : alk. paper) 1. Communication of technical information. 2. Cyberspace. 3. Social media. 4. Distance education. I. St. Amant, Kirk, 1970- II. Sapienza, Filipp, 1966-
T10.5.C85 2010
601'. 4--dc22

2010001620

ISBN 978-0-89503-398-7 (hbk)
ISBN 978-0-415-40318-4 (pbk)

Dedication

For my daughters, Lily and Isabelle, who are a source of inspiration in all that I do, for my wife Dori, whose continued patience and support were essential to this project, and for my grandparents, Lilly and Albert St.Amant, who are the source of my interests in culture, language, and communication.

– Kirk St.Amant

For my family, Paula Powell Sapienza and Hannah Sapienza, for my mother, Vicky Sapienza, and for my grandparents, Filipp and Peggy Kriessl.

– Filipp Sapienza

p

Table of Contents

Culture, Cyberspace, and the New Challenges for Technical Communicators

Kirk St.Amant

Historically, cross-cultural communication was considered an interesting but isolated interaction reserved for individuals who traveled abroad for business or pleasure. As new communication technologies (e.g., film, the telephone, and television) emerged, a growing number of individuals found they could access materials—if not actual persons—from other cultures. The costs and the unidirectional nature of these media, however, still imposed significant restrictions on cross-cultural interactions. It was not until the global diffusion of online media that a significant number of individuals could *interact* on a truly international level. Within this context, Web sites allowed organizations to market their products instantly and cheaply to a range of overseas consumers. Similarly, media such as e-mail and online chat rooms allowed individuals to exchange information, ideas, and materials quickly and directly with international counterparts. At the same time, growing interest in free markets and trade agreements provided incentives for using such media to interact on an international level. Thus, the age of online communication also became the era of truly global discourse.

For individuals working in different communication-based fields, this access to global markets has important implications. Perhaps the most significant of these is the perception of *audience*. Traditionally, when one designed communication products (e.g., instructions), the author or developer thought of audience in terms of local, regional, or perhaps even national terms. Today, however, communication products—particularly those designed for online delivery—are often part of a global process designed to share information and capture market share at an international level. As a result, online communication practices almost inherently involve processes such as translation and localization. And as international online access continues to grow, the notion of communicating with audiences from other cultures will only increase in importance.

ACCESS AND USE IN GLOBAL CONTEXTS

According to *Internet World Stats,* just over one and a half billion persons have online access ("World Internet Users," 2008). While that number represents less than a sixth of the world's population, what is surprising is how quickly such access has grown in the last decade. China, for example, went from just over 22 million individuals with online access in 2000 to over 250 million individuals with access today ("Asia Internet Usage," 2008). India also saw online access skyrocket from 5 million persons with online access in 2000 to 60 million with online access today ("Asia Internet Usage," 2008). Internet use in Africa has also grown markedly—by some 400%—since 2000 with the largest numbers of users in South Africa, Egypt, Morocco, and Kenya (Burns, 2006). Rapid growth is also taking place in much shorter timeframes. For example, the number of Brazilians with online access grew by over 5 million between September 2007 and September 2008 alone, and the number of individuals with online access in France, Germany, and Spain increased by 4 million, 3 million, and 2 million— respectively—in that same period (Burns, 2007a, 2009).

This increase in international online access introduces cultural factors that affect how individuals use online media. In certain instances, linguistic preferences have allowed relative newcomers to beat out better-known companies in certain markets. It was language, for example, that allowed the relatively new Chinese search engine Baidu to beat out the more established Google in order to capture 60% of the search market in China (Grehan, 2008). Other, nonlinguistic factors also have important implications for the use of online media to share information globally. Recent research notes that individuals in Latin America tend to do more online searching than do individuals in the U.S. (Burns, 2007b). The same research also indicates that Internet users in Latin America, on average, spend more time online than do other Internet users around the globe (Burns, 2007b). Additionally, online marketing research indicates that cultures may look for different features—and thus expect different kinds of information—when shopping for the same product online (Chen, 2008).

Within this global online context, appearances can be deceiving. To begin with, raw population statistics might not be an accurate indicator of Internet use. Brazilians, for example, represent the largest online cultural group in Latin America. Interestingly, Chile, which has a much lower rate of online access (one-third of that of Brazil), has over four times the Internet penetration of Brazil (Burns, 2007b). Thus, Chile could actually be a better market for certain online goods and services than Brazil with its larger online population. Similarly, linguistic data can lead to faulty assumptions related to online communication preferences. As a recent study by Yahoo Telemundo and Simmons Research notes, while many U.S.-based Hispanics prefer Spanish-language television stations as a source of information, these same individuals also tend to do their online searching in English (Nelson, 2007). In fact, many of the subjects in the study reported seeking information from Spanish-language television programming while simultaneously searching the Web in English (Nelson, 2007). Thus, effective international communication via online media requires one to go beyond initial ideas related to population sizes, language, and cultural identity.

Further complicating these situations are the definitions—and related assumptions—that the members of a culture associate with the term *online access.* Most Internet users in industrialized nations, particularly in the United States, use laptop or desktop computers to access the Internet. In many emerging economies, however, the power grids and the telecommunications infrastructure needed to use such hardware are problematic. The majority of the Internet users in these areas thus rely on cell, or mobile, phones to access online media via wireless networks. As a result, regions such as Eastern Europe have almost twice the wireless Internet demand than the United States does (Burns, 2007c). Internet access via hand-held devices, however, brings with it different expectations and limitations related to screen size, text size, and so forth. Thus, what constitutes accessible and usable online information can vary markedly depending on the size of the interfaces used in certain regions.

These factors of language, cultural usage patterns, and media preferences indicate that no single strategy can be used for widespread international communication via online media. At the same time, growing global markets for technical goods mean organizations must understand this new online context in order to succeed in today's highly competitive and increasingly international marketplace (St.Amant, 2006). For these reasons, more research needs to be done in the area of culture, communication, and cyberspace. Such an understanding is of particular importance to individuals working in communication-based fields, for they are perhaps the most affected by these developments. This collection of essays examines how these factors of culture and media are affecting one particular field: technical communication.

CHALLENGES FOR TECHNICAL COMMUNICATORS

For technical communicators, the convergence of online media and globalization creates new and important challenges. One major challenge involves creating online materials that can be used by different cultural audiences. A common practice is for organizations to use localization to adapt the content, design, and architecture of online materials for culturally diverse audiences (Yunker, 2002). However, in an era of increasing interchange among culturally different groups, what does being "culturally diverse" mean on the Internet? In an era of increased transnationalism, what do people of polycultural heritage expect in the way of culturally sensitive and shared Internet communication? Some users may require localized content, while others might prefer to use a preexisting set of global literacies for Web interaction. Within this context, technical communicators need to reconsider translation and localization strategies while also keeping current on international market trends that drive such practices.

A second challenge technical communicators face is determining how cultural and linguistic factors affect international online interactions. Certain authors, for example, note that online media can affect the nature of cross-cultural discourse (Barnett & Sung, 2005; Hermeking, 2005). Moreover, other authors have noted that traditional approaches to studying culture and communication sometimes do not seem well suited to examining cross-cultural online exchanges (Ess & Sudweeks, 2005). As a result, today's technical communicators must determine how the structures of online media (e.g., e-mail, chat rooms, bulletin boards, Web pages) affect international discourse. For example, do successful online interactions across cultures involve or require more than just the use of a common language? Similarly, in which respects does the technological infrastructure of online media shape the cultures that interact within these media?

Recent developments also have major implications for education and training practices in technical communication. In sum, educators must determine whether teaching and training in technical communication needs to be modified to address the increasingly international nature of online interactions. If so, what steps should individual teachers/trainers or overall educational and training organizations take to address this topic effectively? Educators also need to address the increasingly international nature of online education and examine how online classes might become mechanisms for exploring international online discourse. These and other issues need to be examined if technical communicators wish to operate effectively within the continually expanding context of international cyberspace.

Addressing these matters is no easy task. It is, however, important that technical communicators begin to examine these issues now, should they wish to succeed in the international online context of 21st-century markets. The essays in this collection are an initial step toward unlocking the complexities that technical communicators need to examine to better understand cross-cultural communication in cyberspace.

OVERVIEW OF THIS COLLECTION

This edited collection examines how cultural factors influence cyberspace interactions and how online media shape cross-cultural discourse. The contributors to this collection use a range of methods—including case studies, empirical research, and usability studies—to address different parts of this area of inquiry. In so doing, these contributors explore a range of topics (e.g., international outsourcing, open-source software, writing practices, legalities of communication) related to using online media effectively with international audiences. The contributors also provide insights into how cultural factors affect a range of professional technical communication practices such as usability testing, Web site localization, online architecture development, and content management. This range of approaches and topics helps readers better understand the complexities associated with international online communication.

The editors of this collection have organized its 10 chapters into three thematic sections. Each section focuses on a particular area in which cultural factors seem to collide with aspects of online media. Through this three-part structure, the text examines themes related to more general technical communication practices as well as reviews more specific cultural and topical aspects involving online media.

Section I: Theoretical Approaches to Technical Communication in Cyberspace

The book's first major section examines different approaches readers can use to conceptualize and contextualize cross-cultural communication in cyberspace. The four chapters in this section present perspectives and insights on the complexities of international online exchanges. These chapters also provide readers with a relatively broad foundation from which they can further explore how aspects of culture and online media affect one another.

In chapter 1, R. Peter Hunsinger argues that technical communicators should be careful not to rely too heavily on *culture* to guide Web localization practices. Hunsinger's reason for this caution is that traditional definitions of culture are often unstable and even unreliable in online environments. For these reasons, Hunsinger suggests that technical communicators might find it more effective to localize online materials based on a broader set of cultural and noncultural contexts (e.g., legal, economic, and technological) that influence an online audience's behavior. In examining these factors, Hunsinger presents ways in which technical communicators can apply an understanding of these contexts to localize online content more easily.

In chapter 2, Clinton R. Lanier builds upon these ideas of localization by examining how audiences from other cultures can actually help with the process of localizing online content. Lanier first provides an overview of the difficulties related to delivering online content to a global audience. He also notes that because such an audience comprises a variety of cultural-rhetorical demands, online content must be closely localized to ensure that information is disseminated accurately and efficiently. As Lanier explains, such close attention to individual cultural-rhetorical preferences has not previously been possible, due to the vast amount of resources required. Lanier then examines how user-customizable online technology can address this situation by providing technical communicators with tools for creating effective products for international audiences.

The localization approaches put forth by Hunsinger and Lanier reveal how important a deeper understanding of culture is to designing online materials for users from other cultures. In chapter 3, Matthew McCool shifts the discussion to show how an understanding of cultural-rhetorical factors is essential when designing overall information systems. McCool explains how optimizing international information systems depends on a number of rhetorical aspects used to engage a target audience. While time and translation are necessary features for international information systems, McCool notes that they are comparatively superficial features. As he explains, for optimization to occur, system developers need to know much more about their target audience—they need an intercultural theory of mind. To examine this idea, McCool's chapter addresses three key factors:

- A computational theory of mind
- Emergent cultural properties
- Information systems optimization via metal computations and cultural properties

In this analysis, McCool reveals the various levels at which one can use an understanding of cultures to develop effective global information systems.

In chapter 4, Martine Courant Rife explores the complex legal situation created by international access to cyberspace. Rife argues that rhetoric has new relevance for virtual environments because of its ability to assist strategic thinking. Rife uses Kenneth Burke's identification and division to discuss the ways in which basic awareness of international intellectual property (IP) law is crucial to the success of technical communication in an age of global online media. The idea is that laws often shape the communication practices that cultures employ when interacting online. To examine this idea, Rife uses examples of international IP law's influence on technical communication. She then explores the ways in which increased knowledge of these issues will benefit technical communicators by allowing them to add value to their organizations via strategic thinking.

Section II: Online Interactions Between Cultures

The book's second section moves the overall examination from general practices to more specific cases showing how cultural factors can affect online interactions. The three chapters in this section offer theoretical, heuristic, and case study analyses of how members of particular cultures use online media to interact with others. Such a focus lets readers compare and contrast the ways factors of culture and media can affect communication behaviors and preferences within specific cultures. This focus also serves as a foundation from which readers can undertake future research involving online media's effects on cross-cultural exchanges.

This section begins with Daniel D. Ding's analysis of how different cultural groups identify, or name, online information. In particular, Ding, in chapter 5, uses the Confucian concept of *naming* to explore the ways in which Chinese as compared with U.S. Web designers organize online content. To do this, Ding compares the organization of Web sites designed by counterpart Chinese and U.S. government agencies. Through this comparison, Ding reveals how different cultural traditions can influence the presentation and organization of online information.

In chapter 6, Carol M. Barnum further explores these ideas by examining rhetorical differences in communication styles between Eastern and Western Web writers. Barnum applies the work of Edward T. Hall and Geert Hofstede to review examples of international business correspondence exchanged via online media. Barnum uses these communiqués to show how aspects of culture and media can contribute to misunderstanding. She also offers suggestions for ways in which analyses such as those presented in her chapter can be useful to teachers and trainers in technical communication.

In chapter 7, "Meeting Each Other Online," a team of international authors reviews a corpus of international online exchanges to explore how the process of collaboration is affected by factors of culture and media. As the authors explain, international online interactions raise issues of language, culture, and the negotiation of common understandings about collaborative tasks. Within this context, one of the first events to take place among members of virtual teams is self-introduction—a process that is important in terms of first impressions. To explore such situations from an international online perspective, the authors analyze a small corpus of online Chinese-U.S. student letters of self-introduction. Through this approach, the authors identify features that index cultural habits and expectations that are important for understanding professional writing practices in international online settings.

Section III: Cross-Cultural Collaborations and Learning Environments

The third and final section moves the focus from that of specific cultures to that of a specific topic—collaborative approaches related to education. The three chapters in this section provide examples of the ways in which technologies are adapted for culturally diverse groups. These chapters also explore how such groups use online technologies to collaborate on learning-intensive projects. This topical focus provides readers with important insights not only for technical communication educators and trainers, but for any individual responsible for developing instructional material for online delivery (i.e., technical communicators in general). The ideas presented in this section also have important perspectives to offer individuals who hire, manage, or train graduates of technical communication programs.

Chapter 8, the first chapter in this section, provides an analysis of how existing instructional and communication technologies can bridge geographic, digital, and cultural divides and facilitate collaboration on an international scale. Through this analysis, the authors—Audrey Bennett, Ron Eglash, and Mukkai Krishnamoorthy—highlight existing Web-based technologies related to distance learning. They also explore the limitations of such technologies for collaboration between first- and third-world participants. In order to do this, the authors use three disciplinary perspectives—those of visual communication, anthropology, and computer science. By combining these perspectives, the authors offer technical communicators an interdisciplinary understanding of important intercultural and technical issues involved in online learning.

In chapter 9, Judith B. Strother reviews the ways in which cultural implications must be considered during all phases of developing Web-based training programs. Such considerations, Strother explains, must address curriculum planning, interface designing, and course delivery involving online and on-site options. To examine these factors, Strother uses a case study of a Web-based English for specific purposes course. Through this case study, Strother reveals the complex cultural aspects that can influence the design and the international delivery of online training.

Chapter 10, by Sipai Klein and Sharon Trujillo Lalla, discusses the ways in which the concept of *digital ecologies* can help individuals learn more about intercultural interactions in learning management systems. Klein and Lalla begin by asking if technologies such as learning management systems meet the needs of globally diverse workplaces and classrooms. To answer this question, they use intercultural variables from Edward T. Hall, Geert Hofstede, and Charles Hampden-Turner and Fons Trompenaars to analyze online learning management systems. Klein and Lalla then use this approach to identify the affordances and the disadvantages of learning management used in intercultural online interactions. Their observations provide valuable insights that technical communicators and technical trainers can use when designing online training for different cultures.

PERSPECTIVES FOR FUTURE ACTION

The area of international online communication is vast and complex. The chapters in this collection represent a small but an important initial step toward understanding the factors contributing to this complexity. For this reason, readers should consider the ideas presented in these chapters as a foundation from which future research can be launched, new educational approaches can be tried, and new professional practices can be developed and tested. Thus, this collection can be viewed as the start of a broader discussion on how the field of technical communication should operate within the international online context of this new century. It is the hope of the editors that readers will view themselves as participants in this process, will further explore the ideas covered in this text, and will carry the overall conversation on to new levels, topics, and cultures.

REFERENCES

Asia Internet usage. (2008). Retrieved January 12, 2009, from the Internet World Stats Web site: http://www.internetworldstats.com/stats3.htm#asia

Barnett, G. A., & Sung, E. (2005). Culture and the structure of the international hyperlink network. *Journal of Computer-Mediated Communication, 11*(1). Retrieved September 10, 2009, from http://jcmc.indiana.edu/vol11/issue1/barnett.html

Burns, E. (2006). Web usage climbs in Africa. *ClickZ.* Retrieved January 10, 2009, from http://www.clickz.com/3603526

Burns, E. (2007a). Active home Internet users by country, November 2007. *ClickZ.* Retrieved December 27, 2008, from http://www.clickz.com/3627971

Burns, E. (2007b). Internet usage in Latin America, June 2007. *ClickZ.* Retrieved October 10, 2008, from http://www.clickz.com/3626562

Burns, E. (2007c). Mobile content usage is higher in developing countries. *ClickZ.* Retrieved October 10, 2008, from http://www.clickz.com/3625143

Burns, E. (2009). Active home Internet users by country, October 2008. *ClickZ.* Retrieved January 10, 2009, from http://www.clickz.com/3632231

Chen, V. (2008). Targeting with culture in mind. *ClickZ.* Retrieved October 10, 2008, from http://www.clickz.com/3629047

Ess, C., & Sudweeks, F. (2005). Culture and computer-mediated communication: Toward new understandings. *Journal of Computer-Mediated Communication, 11*(1). Retrieved September 10, 2008, from http://jcmc.indiana.edu/vol11/issue1/ess.html

Grehan, M. (2008). Gearing up for pan-Asian search marketing. *ClickZ.* Retrieved October 10, 2008, from http://clickz.com/3629914

Hermeking, M. (2005). Culture and Internet consumption: Contributions from cross-cultural marketing and advertising research. *Journal of Computer-Mediated Communication, 11*(1). Retrieved September 10, 2008, from http://jcmc.indiana.edu/vol11/issue1/hermeking.html

Nelson, M. G. (2007). Yahoo study shows Hispanics online are savvy multitaskers. *ClickZ.* Retrieved October 10, 2008, from http://cwww.clickz.com/3625375

St.Amant, K. (2006). Open source and outsourcing: A perspective on software use and professional practices related to international outsourcing activities. In H. S. Kehal & V. P. Singh (Eds.), *Outsourcing and offshoring in the 21st century: A socio-economic perspective* (pp. 229-247). Hershey, PA: IGI Global.

World Internet users and population stats. (2008). Retrieved January 10, 2009, from the Internet World Stats Web site: http://www.internetworldstats.com/stats.htm

Yunker, J. (2002). *Beyond borders: Web globalization strategies.* Indianapolis, IN: New Riders.

SECTION I

Theoretical Approaches to Technical Communication in Cyberspace

CHAPTER 1

Using Global Contexts to Localize Online Content for International Audiences

R. Peter Hunsinger

CHAPTER OVERVIEW

To effectively design international online content, technical communicators often localize materials for culturally diverse target audiences. However, this chapter argues that technical communicators should be careful not to rely too heavily on *culture* to guide localization, since traditionally defined culture is often unstable and unreliable in online environments. Rather, technical communicators may find it more effective to localize based on a broader set of cultural and noncultural contexts (e.g., legal, economic, technological) that influence an online audience's sense of locality. This chapter offers specific ways in which technical communicators may apply an understanding of these contexts to the localizing of online content.

As the Internet develops into a truly world-wide Web, audience expectations about the amount, presentation, and type of online content often become widely divergent, and criteria governing the credibility and usability of content similarly vary (St.Amant, 2002a, 2005b; Yli-Jokipii, 2001; Zahedi, Van Pelt, & Song, 2001). Technical communicators and Web designers attempt to reach the Internet's increasingly diverse audiences by developing sophisticated internationalization or globalization strategies. Using these strategies, Web designers develop online content to make it usable and accessible to multiple international audiences (Cyr & Trevor-Smith, 2004) through, for example, making online content available in multiple languages (translation), selecting software and coding protocols to meet international and region-specific standards, and adapting online content for multiple regional target audiences.

This chapter focuses on the last-mentioned component, usually called *localization*, which involves adapting a single set of content differently to "culturally diverse audiences" (Ulijn & Campbell, 2001, p. 78) or to the "local culture" of a specific target audience (Yli-Jokipii, 2001, p. 105). While I contend that the premise of localizing online content is sound and reasonable, I argue in this chapter that technical communicators and designers should be careful not to rely too heavily on the term *culture* to conceptualize, orient, and guide their online localization strategies. Rather, I suggest that understanding the ways international online audiences construct a more general sense of locality—in which audiences inflect or even modify their cultural habits in light of such factors as political structures, legal frameworks, historical contexts, economic concerns, or scope of community—is a safer and more pragmatic approach to localizing online content.

As the first two sections of this chapter will clarify, such a shift away from the traditional focus on culture is necessary because "culture" is an unreliable guide in dynamic, globalized online environments. In such spaces, the intense circulation of radically different values, norms, and ways of life requires online users to renegotiate traditional cultural habits, and culture tends to function more as a means of maintaining stability than as a description of predefined habits and values. The next two sections of this chapter, then, will sketch an approach to localization in which cultural and noncultural global contexts intersect online to shape a target audience's constructed sense of locality. I will conclude the chapter by offering an analytical tool that may help technical communicators and Web designers identify and use these global contexts to localize their online content, including an example application of the tool to Philip Morris Inc.'s localized U.S. and international sites.

THE FUNCTION OF CULTURE IN ONLINE LOCALIZATION RESEARCH

To clearly define an online localization strategy that relies on a target audience's constructed sense of locality rather than its culture, let's begin by considering the ways in which culture has often served as an orienting and guiding term for online localization efforts, as well as the results of this emphasis on culture. I call this type of localization strategy the culture-based approach.

Culture, I argue, tends to become a guiding term for online localization efforts when it serves to define a target audience and name the audience's salient perspectives and practices, in other words, when the target audience's culture is the primary characteristic that Web authors and designers rely on to localize online content. For example, Yli-Jokipii (2001) observed that a Finnish company he studied used "cultural stereotypes" (p. 110) to define its Finnish- and

English-speaking audiences and to tailor the online content on their respective sites. He explains that the Finnish-language site contains minimal information, which suggests a culture in which "the role of . . . the information contained in the context is quite significant" (p. 110). The site, moreover, supports the cultural stereotype of the "silent Finn," who has a relatively "high tolerance of silence" and who expects much information to remain unsaid. The English-language site, by contrast, contains profuse, detailed information, which evokes a cultural system that prefers not to rely on shared tacit knowledge but rather on explicit information.

More generally, St.Amant (2005b) has suggested that Web authors and designers should vary the specific images displayed on a Web site, the colors used, and the persuasive strategies deployed depending on the cultural practices and perspectives of the target audiences (pp. 76-77), and Zahedi et al. (2001) differentiated target audiences by "national cultures" (p. 85), for example, German, Swedish, and so forth, to discuss methods for localizing online content (also see St.Amant, 2002b, p. 200; Ulijn, Lincke, & Karakaya, 2001, p. 131). Moreover, some researchers have refined or broadened their use of culture by using regional, religious, or ethnic adjectives, referring, for example, to Islamic cultures (St.Amant, 2005b) or Nordic cultures (Yli-Jokipii, 2001). The common thread among these examples is that the term culture carries most of the weight of defining online audiences and localizing content.

When localization efforts thus rely on culture as their guiding term, the result is that traditionally noncultural contexts of web use (e.g., political, legal, historical, economic, technological) become pushed to the background. In part, this is because several characteristics that define an audience and its context of Web use tend to be grouped under the category of culture. Yli-Jokipii (2001), for instance, describes the differences between a Finnish company's Finnish- and English-language sites in terms of the target audience's cultural values and expectations. While the author keenly identifies important divergences between the two versions, and while some of these divergences are certainly cultural, important local characteristics, such as the users' national contexts or economic purposes, may help account for the sites' differences more specifically. National contexts, for example, can imply traditionally noncultural characteristics such as legal frameworks and political structures, and Yli-Jokipii (2001) notes that the sites are, in part, localized around economic purposes rather than culture: retail information for Finnish-speaking audiences and investment information for English-speaking audiences. Similarly, St.Amant (2005b) uses the example of wall outlets to demonstrate how seemingly universal images in fact vary by locality (e.g., two flat prongs in the United States, three prongs in Australia). This knowledge is certainly crucial for localization strategies and helpful for designers, and St.Amant's (2005b) characterization is by no means misguided, but, while wall outlet shape depends quite a bit on traditionally noncultural factors such as the audience's local political jurisdiction and technical regulatory authorities, the author describes the variation in terms of "cultural expectations of what features an item . . . should possess" (p. 76).

In part, the heavy emphasis on culture also pushes noncultural contexts into the background because of the way localization researchers define culture. That is, as Hewling (2004) argues, researchers studying culture in online environments tend to draw their premises and definitions from anthropology and workplace communication studies, and these disciplines have typically treated culture as a relatively stable and static determinant of social behavior, one that often functions with little regard for situational context. In technical communication research, for example, researchers studying online localization strategies have often relied on Geert Hofstede's (1984) intercultural workplace communication research (St.Amant, 2005a, pp. 141–142; Ulijn & Campbell, 2001, p. 78; Yli-Jokipii, 2001, p. 110; Zahedi et al., 2001, p. 85). Hofstede (1984) described several cultural dimensions, to identify specific areas in which cultures may vary. For instance, in the dimension of power distance, cultures rank on a continuum from low to high: High indicates strict social hierarchies, while low indicates more egalitarian arrangements. While these dimensions have proven useful for systematizing cultural variation and planning some features of online localization strategies, I have argued elsewhere that codifying a cultural system tends to portray an audience's practices, preferences, and expectations as abstracted, static, and context independent (Hunsinger, 2006). As a result, important contexts that may influence a target audience's online behavior (but that fall outside the category of culture) become deemphasized, and the important negotiation and mobilization of culture in online environments becomes invisible. Instead, with Hofstede's (1984) image of culture as a starting point for a localization strategy, a target audience's culture functions as an entity that determines the audience's online preferences, behaviors, standards, and expectations.

GLOBAL ONLINE ENVIRONMENTS AND THEIR CHALLENGE TO STABLE CULTURE

If an audience's culture were stable, static, and context independent, then localization efforts would have no trouble relying on culture as an orienting term, and the culture-based approach would be sufficient. However, culture is often more complex and slippery, and this, as we will see below, diminishes culture's usefulness as a guiding term for online localization strategies and thrusts contextual factors into the foreground. Especially in the dynamism and interactivity of online environments, effective localization strategies must reflect the fragile fluidity of culture and account for individuals' capacities to question or alter their cultural habits or adapt to new ones.

Perhaps most apparent, the constraints of the online communication media themselves contribute to culture's fragility in Web environments, in that online users must often negotiate or reconsider familiar cultural norms in certain cross-cultural online contexts. For example, St.Amant (2002b) has noted that "the plasticity of online identity seems to contradict the communication norms that members of certain cultures use to govern communication practices in terms of how they should behave and to whom they should listen" (p. 202). Elsewhere, St.Amant (2005a) has also observed that "formality breaks down online" (p. 141), explaining that the inherent limitations of non-face-to-face online communication hinder a reliance on familiar cultural norms that may apply in other contexts (also see Zahedi et al., 2001). Moreover, because online media often require specific and novel ways of communicating, largely through text and images in blogs, discussion forums, or e-mail messages, Chase, MacFayden, Reeder, and Roche (2002) have noted that ad hoc norms develop in online environments, including new "rules of formality/informality, flexibility, [and] interaction style" (Chase et al., 2002, Table 2 "Online Culture").

Perhaps less explicitly, however, the very scope of the Internet, as a central vehicle of globalization, complicates and destabilizes traditional concepts of culture. That is, as anthropologist Arjun Appadurai (1996) explains, global communication technologies such as the Internet in effect expose culture to the hyperconnectivity of the globalizing world. As Appadurai (1996) observes, global communications media

> compel the transformation of everyday discourse. . . . They allow scripts for possible lives to be imbricated with the glamour of film stars and fantastic film plots and yet also be tied to the plausibility of news shows, documentaries, and other black-and-white forms of telemediation and printed text. (pp. 3–4)

In other words, "electronic media provide resources for self-imagining as an everyday social project" (Appadurai, 1996, p. 4), offering a multitude of texts with which individuals can identify, to which individuals can adapt, or against which individuals can oppose themselves. The result for culture in online environments is that, as Ulijn and Campbell (2001) observe, "new communication technologies have pushed culture from its normal tacit state into the foreground" (p. 78), thrusting culture into individuals' conscious and intentional social actions.

Sapienza's (2001) concept of "translocal communication" (p. 437) more specifically describes some important ways in which individuals in online environments negotiate, draw from, or reject these circulating texts and images to construct a stable cultural identity. The term *translocal* articulates the ways in which Web designers "may incorporate ideas situated not only across the globe but from the next town, state, or province" (p. 437), taking advantage of the Internet's hyperconnectivity and global scope. In articulating a theory of translocal communication, Sapienza studied 30 Web sites designed to help Russian-speaking immigrants in North America adapt to their new social environments and maintain ties to their homeland. These sites offered technical assistance (e.g., information about local events, religious services, advertising, employment listings) and cultural and artistic resources to the sites' users, who included individuals born or currently living in the United States, Russia or the former Soviet republics, and other locales.

Sapienza's (2001) study elucidates at least two important features of the negotiation of online cultural identity that challenge traditional notions of stable, static culture. First, the study illustrates that the circulating texts and images that Appadurai (1996) described can exist side-by-side, rather than mutually exclusively, to offer multiple possible perspectives for discussing the issues that the immigrant communities may face; in other words, various cultural perspectives and practices are mixed and matched in different situations. Sapienza (2001) explains that "Immigrant web sites do not reflect cultural polarization but rather varying degrees of juxtaposition and mixing of local and global" (p. 435), because the sites he observed did not attempt to divide Russian perspectives from non-Russian perspectives by strictly linking specific practices and perspectives to predefined cultures. Rather, the circulating texts and images that reflected traditionally Russian or North American political, legal, religious, economic, and other identities and frameworks converged within the sites and offered varying perspectives from which to encounter the immigrant experience.

Second, because the perspectives and practices available on the immigrant sites were not mutually exclusive or culturally polarized, the construction of cultural identity among the sites' users relied more on conscious negotiation and identification than on pinpointing an "authentic" cultural identity, whether defined geographically or ethnically. Sapienza (2001) explains that "Russian ethnicity [on the sites] seems to be a less important factor in terms of validating a participant's role in a Russian immigrant community online than it may be in a geographical setting" (p. 437). Instead, the agile negotiation of various perspectives, as well as the content contributors' familiarity with Russian and non-Russian cultural texts and practices, validated participation. Moreover, Sapienza (2001) describes the emphasis of the sites on negotiating various perspectives, as users experienced "the new culture through the prism of the old, giving them a reprieve from the stress of new surroundings, as well as a homeland context from which to interpret them" (p. 441). That is, the users drew from different experiential frameworks available on the sites and in their lives to define and negotiate an individualistic immigrant identity.

While these translocal practices that Sapienza (2001) described are especially vivid on the immigrant Web sites he studied, we should consider these practices to be a general feature of negotiating identity and usability in online environments. Sapienza (2001) argues that "knowledge of translocal communicative practices will greatly facilitate document designers as they . . . encounter general audiences with greater exposure to global cultures" (p. 446), but this knowledge can prove useful for understanding even apparently culturally polarized situations. If we return to Yli-Jokipii's (2001) study, for example, we can see that the "cultural stereotypes" the author identifies can be thought of in terms of different interacting perspectives and horizons of experience in the globalizing world. That is, English-speaking users of the site described by Yli-Jokipii may begin to recognize that they are, in fact, visiting a Finnish site with an identifiable set of perspectives, assumptions, and usability practices, while they also begin to recognize their own practices and usability needs. The company's English-language site, then, offers specific texts, images, practices, and perspectives to help the English-speaking audiences navigate the online content.

CONSTRUCTING LOCALITY IN ONLINE LOCALIZATION STRATEGIES

Within global online environments, which require individuals to negotiate a multitude of texts, images, practices, and perspectives, Web designers cannot easily rely on traditional concepts of stable and static culture to develop online localization strategies. However, culture does not simply become irrelevant online, and understanding a target audience's practices and preferences (traditionally cultural concerns) remains crucial for effective localization. Rather, cultural concerns must be contextualized as a specific element among the contexts of the globalizing world. As I argue in this section, understanding online users' constructed sense of locality, which includes culture as an irreducible component, may guide localization strategies more effectively than relying on traditional concepts of culture. I call this type of localization strategy the locality-based approach.

Appadurai (1996) describes an individual's or group's sense of locality as "a complex phenomenological quality, constituted by a series of links between the sense of social immediacy, the technologies of interactivity, and the relativity of contexts" (p. 178). These technologies (which expand or limit the scope of locality) and contexts (in which individuals and groups distinguish self from other, inside from outside) form the "social, material, [and] environmental" "ground" against which locality is "imagined, produced, and maintained" (p. 184). In other words, a sense of locality, rather than simply being the given result of a physical environment or a material community, emerges from the active social work of determining in from out, us from them, or here from there.

It is this social work necessary to produce locality that is crucial to understand in global online environments. Online, the production of locality involves creating a sense of delimited community or identity within a broader "ground" or context marked by the global hyperconnectivity of online communication technologies and a vast network of texts, images, interpretive frameworks, and horizons of experience. For instance, designers can create a sense of locality in online environments through password protection or site registration, which limits site access and distinguishes in from out; designers may also define relationships to other sites and communities (links and references), which trace an online territory. Similarly, users can develop their own sense of locality by selecting sites to frequent (walking an online territory) and by choosing to interact with certain personae in online discussions (distinguishing us from them). Generally, these delineations serve to define a specific (if amorphous) group of users, who share similar frames of reference, within a specific territory of connected Web pages.

Within these global online environments, the active, symbolic maintenance of locality is especially apparent when we consider that individuals rely more on their familiarity with various images, texts, practices, and perspectives than on traditional markers of geography, ethnicity, or predefined culture to circumscribe locality. For example, the online Russian immigrant communities that Sapienza (2001) described define a sense of locality by linking to sites with appropriate technical or cultural content, but the selection of links is not based on geography or ethnicity. Sapienza explains: "The criterion for 'valid' linking to Russian-related material . . . rests on the content of the linked pages, wherever they happen to be located" (p. 441). Establishing online locality thus relies heavily on the practices of the community's participants, rather than on the predefined identities or physical locations of the participants.

However, as suggested by the emphasis that the Russian immigrant sites place on the content of the linked pages, the idea of a culture or cultural identity still plays an important role in defining locality in online environments. That is, as noted above, while the sites that Sapienza (2001) described do not reflect cultural polarization (relying instead on an interplay of cultural perspectives and practices), the idea of culture still exists. It seems clear that, while some content on these sites is deemed appropriate because of its practical value (e.g., legal information or material that covers immigrant adjustment issues), some criteria exist that allow users to determine the appropriateness of the content that attempts to educate nonimmigrant users in Russian cultural forms (Sapienza, 2001, pp. 440–441). These latter criteria, I contend, seem to rely on the idea of a Russian culture to determine appropriateness. However, the *use* of a culture as a vetting criterion should be distinguished from the existence of a stable, static culture. That is, as Appadurai (1996) explains, culture in global online environments is becoming less a name for a collection of stable and static perspectives and practices and more "an arena for conscious choice, justification, and representation" (p. 44). Culture, in other words, is increasingly a self-conscious rallying point for the mobilization of group identities, a means of stabilizing identity and creating a sense of locality within an unimaginably vast social space. In the case of the Russian immigrant communities, Russian culture functions more as a distinguishing criterion for certain content than a description of a preexisting Russian-ness.

To give an example, as Kitalong and Kitalong (2000) have described, some indigenous inhabitants of Palau, following their independence in 1994, were concerned that "inaccurate materials posted about Palau on the Internet could be construed as the truth about Palau and its culture" (p. 105). Palauan Web authors then developed their own online content to describe a national identity that challenged the common Western perceptions of Palau (as either a tourist attraction or a World War II battleground) and a cultural identity that challenged caricatures of the nation's indigenous peoples (as one-dimensional, tropical exotics). In Appadurai's (1996) terms, Palauan Web authors entered the arena of global media, which circulated the imagined lives and flat representations of Palau and its people, and self-consciously mobilized Palauan culture online to assert a unique and independent identity. Moreover, the Palauan Web authors circumscribed a sense of locality, in part, by distinguishing "authentic" from "inauthentic" representations of Palau and its peoples and by dividing self from other: one Palauan politician, for instance, was concerned that Palauan "values could be corroded from exposure to other cultures" (Kitalong & Kitalong, 2000, p. 110). But Palauans also developed a unified conceptual map of Palau's geographical, ethnic, and cultural features (p. 105), all against the innumerable possibilities of Palauan representation circulating in global online environments.

This example from Kitalong and Kitalong (2000) and Sapienza's (2001) description of the Russian immigrant communities also show that understanding the construction of locality requires understanding the convergence and interaction of a number of contextual factors (technological, ideological, geographical, economic, etc.). These contexts, which mainly consist of traditionally noncultural factors, influence an individual's sense of locality and identity online. For example, accelerating waves of Russian immigration to North America constituted the exigency for the Russian immigrant sites, and the sites defined a sense of locality not only against the texts and images of global online environments but also against a geographically dispersed network of users; the circulation of stereotypical images of Palauan identity over global communications media moved the Palauan Web authors to mobilize the idea of Palauan identity, thus asserting a postcolonial locality online; and the economic reach of Western tourists and Western colonial powers created the sense of Western encroachment, against which Palauan identity was mobilized.

Recognizing that stable, static culture does not solely and ultimately describe international online communication, Zahedi, Van Pelt, and Song (2001) developed a framework for characterizing a target audience and some of these contextual factors, using age, gender, economic, and other demographics, as well as users' experience with information technology and habits of information processing (p. 85). However, Appadurai (1996) offers a more comprehensive framework that can also help identify the systematic patterns and interactions of these various contexts. He organizes these contextual elements in terms of five scapes, or different material and textual fields, flows, and patterns that converge and overlap to form an individual's sense of locality or identity. Appadurai (1996) characterizes the scapes as "deeply perspectival constructs" (p. 33), though they have material (textual or otherwise physical) existence, which individuals negotiate to organize and characterize their experiences: These scapes form the elements of an individual's sense of, for example, German-ness or Thai-ness. Appadurai (1996) describes the five scapes as follows:

- *Ethnoscapes* describe the "landscapes of persons who constitute the shifting world in which we live: tourists, immigrants, refugees, exiles, and other moving groups" (p. 33), all of which bring widely divergent ways of life into close contact with one another. Thus, ethnoscapes also describe what has traditionally been called culture, the embodied practices, preferences, and perspectives that circulating populations carry with them. These movements of peoples and practices occur in patterns that affect different peoples and areas of the globalizing world differently: for instance, while many immigrants today tend to move from south to north, or from developing to industrialized areas, many tourists tend to move in the opposite direction. Such patterns, for example, formed the backdrop against which the Palauan Web authors and the Russian immigrant communities developed their online content.
- *Technoscapes* describe "the global [and differential] configuration . . . of technology" (p. 34). Most salient for understanding online locality are information and communication technologies, such as the Internet, that are accessible to different degrees in most areas of the world and constitute the "technologies of interactivity" (p. 178) that influence the scope of locality. For example, the Internet allowed the Russian immigrant sites (Sapienza 2001) to develop a locality that included individuals in North American and the former Soviet republics.
- *Finanscapes* describe individuals' relationships to global capital flows as well as the different economic classes of the world's nations and peoples. While finanscapes can influence both ethnoscapes (e.g., allowing the privilege of tourism for certain groups or driving immigration) and technoscapes (e.g., differentiating access to technologies), finanscapes also influence the relationships between the members and nonmembers of a locality. For example, the Palauan Web authors (Kitalong & Kitalong, 2000) challenged the supposed privileges of the West's global economic position, which had allowed the West to use Palau as the stage for its fantasies.
- *Mediascapes* describe "both the distribution of the economic capabilities to produce and disseminate information" and "the images of the world created by these media" (p. 35), and, like ethnoscapes, mediascapes circulate in patterns influenced by finanscapes and technoscapes. The creation and use of these images are a key component in developing a sense of locality online, as they differentiate "home" from "away" and describe the members included in any particular locality. For example, while Western

media centers developed and circulated images of Palau as an exotic paradise, the Palauan Web authors mobilized the media at their disposal to offer a counter-set of Palauan images.

- *Ideoscapes* describe the circulation of images and texts that are "often directly political and frequently have to do with the ideologies of states" (p. 36), including individuals' relationships to government entities or recognition of the legal legitimacy of certain institutions. In terms of developing a sense of locality online, ideoscapes constitute the various political perspectives and legal frameworks that individuals must negotiate to determine the procedures that a locality adheres to and the jurisdictions under which it operates. For example, the Russian immigrant sites in North America offered informal legal advice to their members in a way that recognized the U.S. system as a legitimate authority.

While Appadurai (1996) uses the scapes to describe an individual's sense of identity or locality, I suggest that Web authors and designers may also use these scapes to characterize their target audiences. For example, let's examine Sloane and Johnstone's (2000) description of the Scottish nationalism prevalent throughout online Scottish news sites. The authors observe that "Critical readers of Scottish newspapers on the Web today are likely to find themselves interacting with online reporting that reveals cross-cultural dynamics of power and national identity as much as it relays the news" (p. 155). In other words, though Scotland is a well-defined geographical entity, Scottish readers must still develop and maintain a sense of Scottish locality, for which nationalism has been perhaps the most obvious banner. The scapes inflect this sense of locality:

- Mediascapes, when "the Web conveys the signs (and distortions) of Scottish identities" (p. 161) that Scottish readers must negotiate
- Ideoscapes, when individuals practice online literacy in the shadow of "an uncomfortable union with a dominant partner in the south (England)" (p. 163)
- Ethnoscapes, when individuals identify themselves as "Scottish," wherever they are around the world, and appear skeptical of "England and English ideas of education, style, literary worth, and critical taste" (pp. 155–156)
- Finanscapes, when the editor of the *Scotsman Online Edition* portrays the paper as a nationalistic economic equalizer: "although the majority of visitors will be those that cannot buy the terrestrial product, everybody is welcome" (pp. 166–167)
- Technoscapes, when the very scope of Scottish locality is challenged, broadened, or made virtual through Internet technologies

Moreover, even in this situation where a rather clear national identity and delineated cultural traditions are in place, note the fragility of Scottish culture and the influence of traditionally noncultural contexts on the shape of Scottish identity online. That is, Scottish culture, instead of describing the practices and perspectives of a people who share Scottish-ness, is rather an orienting ideal around which various practices and perspectives develop into a sense of online Scottish locality.

IMPLICATIONS FOR ONLINE LOCALIZATION METHODOLOGIES

Drawing from the framework for online localization that I have described, this chapter, I will describe ways in which online technical communicators and Web designers may begin to apply the locality-based approach to developing effective online localization strategies.

As I have argued in this chapter, basing online localization strategies on a concept of static and context-independent culture—the culture-based approach—can be problematic because of the complicating influence of online environments: multiple cultural texts, images, perspectives, and practices can exist side by side, rather than mutually exclusively, in online environments; navigating online environments relies more on negotiating different perspectives and practices than on adhering to a single preexisting set of cultural tenets; and culture functions more as a self-conscious mobilizing principle than as a description of preexisting beliefs and practices. As a result, localization strategies that rely heavily on culture, while they may identify many important characteristics of the target audience, also tend to obscure the contextual factors that shape and inflect a target audience's practices and perspectives, and the ways in which individuals negotiate and mobilize cultures become a negligible concern. Thus, the orienting term for localization efforts, whether culture or the audience's sense of locality, makes visible a certain set of Web design strategies that are available for addressing a target audience and conceals other potential strategies.

Moreover, relying on culture to localize online content is especially difficult for Web authors who are not members of the target audience or who do not have the resources for a professional localization service. For example, if an online audience does not consider the site's designer or institution a member of the target audience community, culturally based localization may seem disingenuous or pandering (Cyr & Trevor-Smith, 2004). Indeed, a review of several localization services that appear in a Google search indicated that most services pride themselves on their membership in local communities. For example, Webtraduction (2004) proclaims that "We only work in the languages of the cultures that we live in," and SDL (2008) offers customers "fully out-sourced localization of all global content."

In contrast to relying on the orienting idea of culture, I have argued that understanding the ways individuals develop a sense of locality, in light of traditionally noncultural factors (the scapes), can be more effective for understanding the usability needs of online audiences. To develop specific localization strategies along these lines, a guiding principle is Sapienza's (2001) "translocal communication." Sapienza argues: "Designers may have to contrive ways to adopt translocal literacies that would enable effective communication strategies that are sensitive to particular cultural values yet are flexible to overlapping and interplay among global ideas" (p. 442). In other words, effective online localization understands the ways in which audiences inflect or modify their cultural habits and preferences in light of geographies and professed identities (ethnoscapes); political values or relationships to political and legal structures (ideoscapes); economic needs and relationships to global markets and capital flows (financescapes); access to and habits of using communication technologies (technoscapes); and images of possible lives and horizons of experience (mediascapes).

While many online localization services offer to translate content, adapt text and images to local preferences, and configure software and coding for local protocols, the interactions among these factors and the effects of these interactions on audiences' usability needs and preferences do not seem to be an explicit part of many localization strategies. To take a representative example, MultiLing (2007) offers localization services it calls "100% Correct—Grammatically and Culturally." While there is no reason to doubt MultiLing's expertise, claiming to be 100% culturally correct implies a culture-based approach to localization: the claim suggests that content can be adapted to a preconstituted and strictly circumscribed cultural audience and that audience preferences exist independently of the culturally destabilizing forces of, for example, global political contexts or worldwide communication technologies. In contrast, using the scapes can help designers and researchers identify the "global ideas" circulating online and translate them into a target audience's coherent local experience in global online spaces.

For practical Web design, the scapes can perhaps function most effectively as a planning tool that helps Web authors characterize target audiences and target locales. Web authors planning a localization strategy may apply the following analysis method, which involves questions addressing various design considerations:

- Mediascapes. What local and global images circulate among the target audience and within the local context? What popular media narratives, or "scripts for possible lives" (Appadurai, 1996, p. 3), are deeply ingrained within the local community? How is the institutional persona that runs the site perceived within the local community? Are the site's images and texts appropriate for these popular perceptions, images, and narratives, and does the site address issues or controversies specific to the institution and locality?
- Ideoscapes. What legal and political structures, both local and international, shape the audiences' lives within the local community? What political narratives shape the audiences' perceptions of their locality and of the world?

Do the target audiences perceive these legal structures and political narratives as reliable, accurate, or trustworthy? How might the site's images and texts incorporate or skirt certain political narratives and audience attitudes?

- Ethnoscapes. Though Hofstede's (1984) traditional cultural categories (e.g., power distance, uncertainty avoidance) fall under ethnoscapes, classifying these cultural categories as ethnoscapes emphasizes how cultural habits are shaped by mediascapes, ideoscapes, finanscapes, technoscapes, and demographic factors. Important questions: How do the site's audiences identify themselves within the local community, and what ethnic, racial, and cultural divisions are important for the local community? How do audiences perceive and relate to groups outside their local community, with mistrust, respect, derision, or tolerance? Will male and female and younger and older users access the site in equal numbers and in the same manner, or does Internet access skew toward particular audience segments? How might the site appear sensitive to the needs of minority audiences and carefully ensure full audience representation? How might the site meet culturally ingrained usability habits and create a culturally sensitive interactive experience?
- Finanscapes. Will the localized site operate in an area in which Internet access is widespread and evenly distributed, or will the audience be limited to a wealthier subset of the local community? Is the local community intensely involved in economic globalization, or is international trade relatively new to the local community? What are audience attitudes toward wealth and privilege: Is there emphasis on perceived social equality or is strict hierarchy generally accepted?
- Technoscapes. Do global communication networks traffic heavily through the locality, or is the technological infrastructure relatively sparse? Can most segments of the local audience access the Internet (and the site), or is Internet access limited to particular subsets of the local audience? What Internet technologies, software protocols, or file types are available or preferred within the locality?

Perhaps the clearest way to demonstrate the applicability of the locality-based approach is to connect the characteristics of a target audience, organized around the scapes, to specific design elements, in a sense, reverse-engineering a real-world site that appears to rely more on local contexts than culture to localize its online content. The rest of the chapter will be devoted to this sample application, using a comparison of the localized Philip Morris USA and Philip Morris International Web sites as an example. Both companies are tobacco sales subsidiaries of Altria Group, Inc., which had been called Philip Morris Companies, Inc., until January 2003, and the salient difference between the companies, for the present purposes, is their respective market areas. With this sample application, I hope to illustrate not only the ways in which noncultural factors influence audience usability habits, but also the ways in which locality or local community can be, in some instances, a more helpful guide for localization efforts than culture.

Generally, while traditionally cultural concerns are evident in both of the Philip Morris sites' design strategies, a comparison of the sites suggests that political and legal issues (ideoscapes), as well as public image concerns (mediascapes) have mostly guided the localization efforts. In the U.S. site especially, Philip Morris goes to great lengths to distance itself from the growing antismoking sentiment in the United States, as well as the fallout from the "Big Tobacco" lawsuits of the 1990s. The international site, in contrast, attempts to minimize its use of local images and perspectives (mediascapes, ethnoscapes, ideoscapes) and instead create a sense of cosmopolitan, international locality.

For the U.S. site, a designer may outline the scapes as follows:

- Mediascapes. Over the past couple of decades, information about individual, class action, state, and federal lawsuits against Big Tobacco, which includes Philip Morris, has circulated widely, as have antismoking campaigns from both public and private sources. Though millions of Americans continue to use Philip Morris products, the company has suffered a public image crisis.
- Ideoscapes. The U.S. site targets individuals within the U.S. legal framework and political system. Many individuals, whether citizens or documented or undocumented residents, and whether happy or unhappy with the specific state of affairs, tend to accept political/legal structures and services as stable.
- Ethnoscapes. Most of the U.S. audience speaks either English or Spanish as a first or second language, and most individuals consider the United States a multicultural, multiracial society. This fact makes it increasingly difficult for public image campaigns to favor any ethnic, religious, or racial group over others. In online environments specifically, formerly underrepresented groups, such as women or the elderly, are achieving parity among U.S. online audiences (Center for the Digital Future, 2007).
- Finanscapes. In the United States, low-income groups tend to use tobacco more than others (Bobak, Jha, Nguyen, & Jarvis, 2000), while Internet use is increasing rapidly among those in lower-income brackets, approaching parity with other economic groups (Center for the Digital Future, 2004).
- Technoscapes. The U.S. audience has become well versed in Internet communications technology: According to UCLA's Center for the Digital Future (CDF, 2007), about 78% of Americans used the Internet regularly by 2007, either at home or at work, for an average of 8.9 hours per week.

As we will see below, the specific locality of the audience, including locally circulating images, political/legal jurisdictions, and Internet use demographics, none of which are traditionally cultural concerns, drive the site design. Generally, Philip Morris USA seems to count itself as part of a concerned public, aligning itself with prevalent antismoking campaigns and siding against itself in the Big Tobacco lawsuits.

The Philip Morris USA site seems to respond to the popular negative connotations associated with its name and attempts to reconfigure its public image. Most noticeably, the site employs a busy interface, with three navigation bars and five substantive content boxes. In part, the site seems designed for a Web-savvy audience capable of navigating the site's basic features (a response to local technoscapes). But the design also responds to public image concerns (mediascapes) and deemphasizes information that may implicate Philip Morris USA as a tobacco seller. There are, in fact, only few mentions of "tobacco" or "cigarettes," and the site's subtitle refers to Philip Morris USA as a "responsible, effective, and respected developer, manufacturer, and marketer of consumer products, especially products intended for adults." The international site, in contrast, refers to Philip Morris International as a "leading global tobacco company."

Rather than identifying itself as a tobacco company, Philip Morris USA appears to identify itself as a community center offering health-related advice about smoking (responding to mediascape images of amoral Big Tobacco). For example, while the international site offers a link to "Jobs and Careers" at the center-left of its page, emphasized by the use of white space and bold type, the U.S. site offers a tiny link to "Careers" hidden in the lower right-hand corner of the page, embedded in a list that includes "Community Involvement" and "Youth Smoking Prevention." That is, rather than joining Philip Morris USA in its capacity as a tobacco company, the audience is invited to join the company in its capacity as a community center. Moreover, the site notably drops its historical company logo in favor of a nondescript sans-serif "PhilipMorrisUSA" in the upper right-hand corner, which distances the company from its traditional tobacco products.

The community theme also seems evident in the images on the site, which draw from U.S.-specific mediascapes and ideoscapes to offer images of family (a smiling Caucasian father and son) and consumer research (a dark-skinned hand in a button-down sleeve reviewing a book containing graphs). The family image suggests a caring community program (mediascapes), while the studying hand suggests a specific relationship between the U.S. audience and its legal system, where consumer research and advocacy can accomplish safety and health goals through proper channels (ideoscapes). These images also evoke ethnoscapes, mirroring the racial diversity specific to the U.S. audience, though all of the images are of men, which may be a remnant of the period when women were grossly underrepresented in online environments.

Moreover, ethnoscapes and technoscapes are evident in the site's available languages: the U.S. site offers both English- and Spanish-language pages, reflecting the large Hispanic population in the United States as well as the shrinking gap between Internet use among minorities and overall Internet use. Additionally, because the translated pages are identical in design and imagery, it seems evident that cultural concerns did not primarily drive the site design. Rather, population demographics (ethnoscapes) drove the choice to translate the content using an identical site design.

The Philip Morris International site, in contrast, offers a much more cosmopolitan, international experience than the U.S. site. The site offers translations into all of the world's major languages, but the site design is identical for each language, and, presumably, for each culturally specific audience. The images on the various international pages also add to this cosmopolitan theme: when a user accesses one of the international pages, the site randomly loads one of about eight nondescript images without regard for local language or cultural specificity. For example, the Korean international page and the French page are identical and draw from the same set of front page images (see Figures 1 and 2). The only differences among the international pages are the language used and the telephone and email information under the "Press Center" link. All other contact is routed to Philip Morris International's New York City offices.

A designer for the international page may outline the scapes as follows:

- Mediascapes. Using one "international" site to encompass the non-U.S. world suggests that the audiences likely do not share broad common knowledge or opinions about tobacco-related issues. However, Philip Morris is an internationally recognized name, and audiences probably have at least some knowledge of smoking/health issues, though tobacco use does not always carry the same stigma throughout the world as it does in the United States.
- Ideoscapes. Because audiences access the site from a number of political jurisdictions and legal frameworks, coherent or unified ideoscapes cannot easily be specified. However, audiences for the site are unified by some concern for this multinational company, which suggests knowledge of international arrangements.
- Ethnoscapes. While the audience is irreducibly multicultural, multiethnic, and multinational, usable ethnoscapes begin to emerge in Internet use data. For example, while users in industrialized nations constituted most online audiences a decade ago, St.Amant (2005b) explains that Internet use is growing rapidly in developing nations. Therefore, many cultural and national groups are well represented online. However, the CDF (2004) estimates that the average online gender gap throughout the world is around 8%, though it is much higher in certain localities, such as Italy or Spain (20% gap).

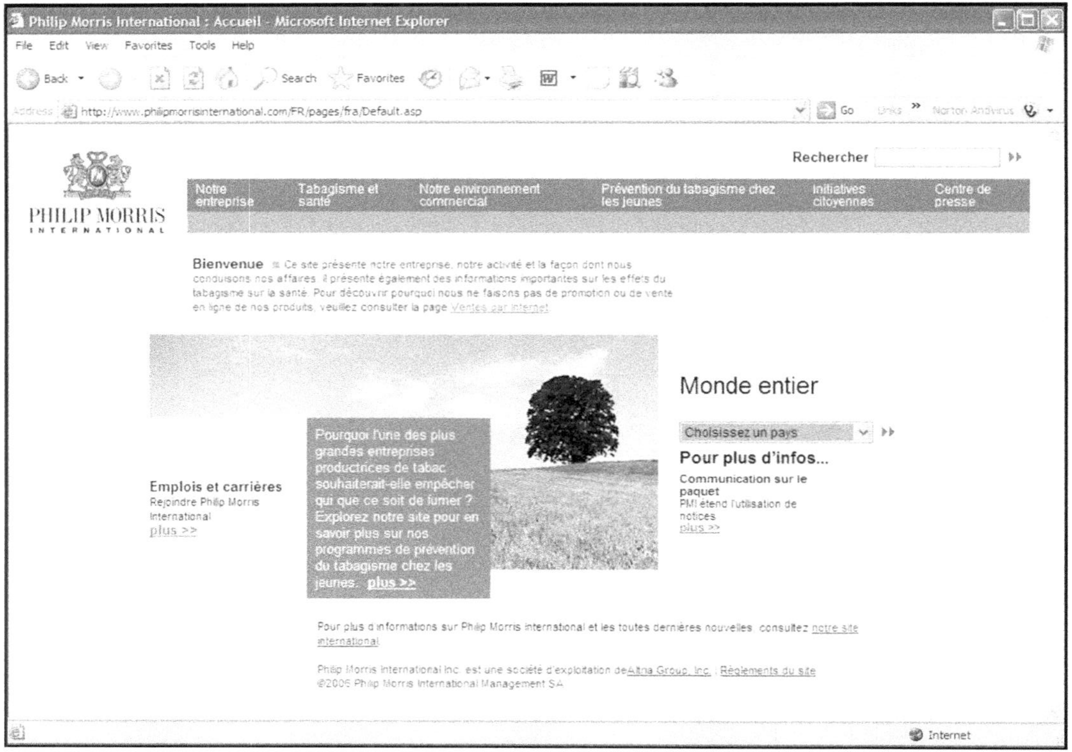

Figure 1. Philip Morris International's homepage, in French (http://www.philipmorrisinternational.com/FR/pages/fra/Default.asp). Courtesy of Philip Morris International.

Figure 2. Philip Morris International's homepage, in Korean (http://www.philipmorrisinternational.com/KR/pages/kor/Default.asp). Courtesy of Philip Morris International.

- Finanscapes. Internet use is low among the poorest quarter of the population in places such as Hungary (1.6%) or Britain (24%), though places such as Korea (46%) or Sweden (49%) are closer to economic parity (CDF, 2004). However, if we cross this data with the finding that tobacco use is especially prevalent among low-income groups (Bobak et al., 2000), we may assume that audiences may not include large numbers of tobacco users.
- Technoscapes. The CDF (2004) estimates that Internet usage rates and habits vary considerably around the world, though it is clear that the audiences live in places that have the technological infrastructure to support Internet use.

Because of the wide variation in audience mediascapes, ideoscapes, and ethnoscapes, the international site seems determined to avoid local specificity, apart from offering the content in several languages (for minimal usability) and press-related contact information. However, avoiding locality is a kind of localization strategy in itself, which draws more from finanscapes and technoscapes than other culture- or region-specific scapes. That is, the international site seems to localize its content for a cosmopolitan, placeless community, and in this way attempts to target a specific audience. The general strategy, then, appears to be to gamble that the audience can accept a sense of cosmopolitan, international community, while Philip Morris International positions itself as a global aid organization or concerned nongovernmental organization (NGO).

As Figures 1 and 2 show, the international site's minimalist design immediately suggests this cosmopolitan theme, with two blocks of content text, a single navigation bar across the top of the page, a generic landscape image at the center, a drop-down menu linking to other localized pages, and a few other minor features. As a whole, the site includes only the basic elements that identify it as a corporate Web site, or, in St.Amant's (2005b) terms, the minimal site "prototype" that audiences can recognize. For example, the site uses short phrases rather than culture-specific icons to indicate links, making the secondary pages more accessible to audiences who may not recognize certain icon conventions.

The lack of icons seems to be part of a broader strategy to minimize imagery. The single central image, devoid of identifiable people and programmed to load randomly from a short list of stored images, attempts to minimize misunderstandings about racial/ethnic tensions, clothing conventions, gender-appropriate dress and behavior, and other culture- or region-specific representations (St.Amant, 2007). The scaled-down design, then, leaves little room for culturally specific content, colors, or design features (avoiding specific mediascapes or ethnoscapes) that may make the site unacceptable to the broad, unmanageable set of audiences. The minimal design also ensures the site's usability in different browser configurations and for audiences that may be relatively new to online environments (based on finanscapes and technoscapes). For example, most of the page is coded in HTML and CSS, with a few lines of

JavaScript to load the page's central image. The site thus avoids relying on browser add-ons that may prevent the page from being usable in certain contexts (St.Amant, 2007).

The site's content also suggests sensitivity to the finanscapes that shape the audiences, who appear to be more concerned about Philip Morris International itself than the products the company sells. The welcome message, for instance, emphasizes the company and the industry rather than the products. Additionally, in contrast to the U.S. site, the international site prominently displays the Philip Morris International logo in the upper left-hand corner of the page, emphasizing it with its positioning, white space, and the contrast of the logo's red and burnished bronze appearing against the site's white field. The site thus draws attention to the company itself, gathering ethos from the brand name or the logo's generally regal look (mediascapes regarding the specific company or official, traditional emblems). Moreover, the content text, including the centered message about smoking prevention, appears to portray Philip Morris International as a health-conscious NGO or international agency. The site does not identify nationally specific policies or legal issues (e.g., settlements or smoking bans), but rather focuses its content on the company's mission and its community programs, such as its programs to prevent youth smoking. The site thus seems to be targeted to investors rather than customers, and it is thus designed around the audience's relationships to global capital flows (finanscapes).

Emphasizing the cosmopolitan theme, all of the pages listed in the drop-down menu are identical, apart from the language used on each page, which appear to be direct translations of the same text. Thus, no special consideration is given to any audience, and no distinction is made among ethnoscapes. The drop-down menu itself, emphasized near the center of the page using spot color, white space, and large text, invites the audience to situate itself within an international community. Because, as Figure 3 shows, the names of the nations listed in the menu are translated into the page's language rather than the nations' native languages (e.g., the French-language page lists nations by their French spellings and the Brazilian page lists nations by their Portuguese spellings), the list seems better suited to display the localities in the international Philip Morris community than to help users find a specific language. That is, the list does not follow the conventions of usable translation links (Yunker, 2002), such as linking to translated content using recognizable national flags or the translated content's native language (e.g., the link from the French page to the Korean translation would be in Korean rather than French). The U.S. site, in contrast, links to its Spanish-language translation in Spanish, emphasizing usability over the display of global community. The international site thus plays off the sense of dispersed community made possible by the technoscapes of Internet communication technologies, creating a sense of locality reminiscent of a "global village."

The international site's central image also evokes cosmopolitanism by offering only generic landscapes and cityscapes (e.g., a well-lit skyline, a tropical

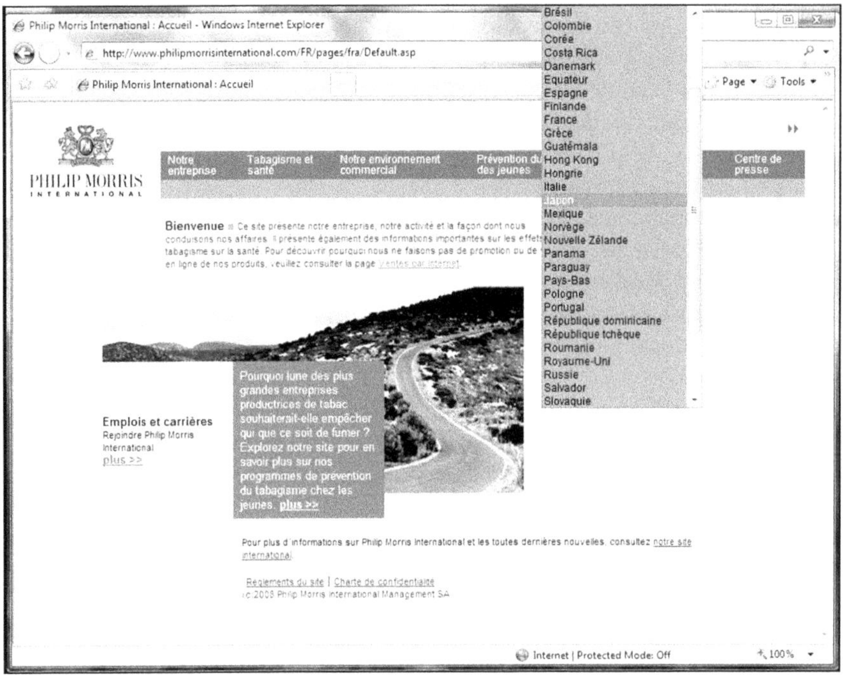

Figure 3. Links to translated content on Philip Morris International's French page (http://www.philipmorrisinternational.com/FR/pages/fra/Default.asp). Courtesy of Philip Morris International.

mountain range, a line of sunny row houses) that may circulate within the audiences' mediascapes. Because the images are not specific to any page, they suggest no specific locality but rather a sense of "somewhere else." That is, the images evoke multinational, cosmopolitan mediascapes that seem designed so that the audiences will not recognize the places as their own. Moreover, as mentioned the site does not depict people, to allow for the varying prototypical images of "men" or "women" (St.Amant, 2005b) and to avoid anchoring the site to a specific place.

In this application of the scapes to reverse-engineering the respective localization strategies of Philip Morris USA and Philip Morris International, we can discern important differences between the culture-based approach to localization and the locality-based approach. The culture-based approach focuses on the specific habits, preferences, beliefs, and perspectives of certain defined cultural groups, for example, identifying the connotations that certain colors or images may have for certain peoples, acceptable strategies of persuasion, or the role that Internet technologies themselves play in certain cultural systems. This knowledge is unarguably important for localizing online content for international audiences and should play a significant part in any effective localization strategy.

However, the locality-based approach can help Web authors and designers plan for other contingencies that arise with international Web design. For instance, this approach can be helpful when many cultural practices and perspectives seem entangled within a particular target audience, or when a single site design must attempt to accommodate numerous diverse audiences, as the Philip Morris International site does. This approach can also help designers plan for the influences that global and local contexts can have on audience habits and preferences, such as the sense of cosmopolitan locality that guided the design of the Philip Morris International site, the public image crisis that guided the design of the Philip Morris USA site, or the technological or financial contexts that guide audience behavior in online environments.

Because understanding and designing around a target audience's sense of locality is heavily context dependent, future research should consider these scapes in light of other localized Web sites with other purposes and within other contexts. Such studies may further refine the scapes' specific applicability for online environments or may identify the horizons and limitations of the context/locality-based approach. Additionally, empirical studies of the effectiveness of this approach may offer designers more specific advice about applying the scapes to localizing online content.

REFERENCES

Appadurai, A. (1996). *Modernity at large: The cultural dimensions of globalization.* Minneapolis: University of Minnesota Press.

Bobak, M., Jha, P., Nguyen, S., & Jarvis, M. (2000). Poverty and smoking [Electronic version]. In P. Jha & F. Chaloupka (Eds.), *Tobacco control in developing countries* (pp. 41–61). New York: Oxford University Press.

Center for the Digital Future (CDF). (2004). First release of findings from the UCLA World Internet Project shows significant "digital gender gap" in many countries. Retrieved June 30, 2008, from http://newsroom.ucla.edu/page.asp?RelNum=4849

Center for the Digital Future (CDF). (2007). Online world as important to Internet users as real world? Retrieved June 30, 2008, from http://www.digitalcenter.org/pdf/2007-Digital-Future-Report-Press-Release-112906.pdf

Chase, M., MacFayden, L., Reeder, K., & Roche, J. (2002). Intercultural challenges in networked learning: Hard technologies meet soft skills. *First Monday, 7*(8). Retrieved October 14, 2006, from http://firstmonday.org/htbin//cgiwrap/bin/ojs/index.php/fm/article/view/975/896

Cyr, D., & Trevor-Smith, H. (2004). Localization of Web design: An empirical comparison of German, Japanese, and U.S. Web site characteristics. *Journal of the American Society for Information Science and Technology, 55*(13), 1–10.

Hewling, A. (2004). "Foregrounding the goblet": Moving on from geographic/nationality based frames of reference when looking at culture in the online classroom. Paper presented at the conference on Cultural Attitudes Towards Technology and Communication, Karlstad, Sweden.

Hofstede, G. (1984). *Culture's consequences: International differences in work-related values.* (Abridged ed.). Beverly Hills, CA: Sage.

Hunsinger, R. P. (2006). Culture and cultural identity in intercultural technical communication. *Technical Communication Quarterly, 15*(1), 31–48.

Kitalong, K. S., & Kitalong, T. (2000). Complicating the tourist gaze: Literacy and the Internet as catalysts for articulating a postcolonial Palauan identity. In G. E. Hawisher & C. L. Selfe (Eds.), *Global literacies and the World Wide Web* (pp. 95–113). New York: Routledge.

MultiLing Corporation. (2007). Web site localization. Retrieved June 28, 2008, from http://www.multiling.com/Web.aspx

Philip Morris International Inc. (2007). Home. Retrieved October 14, 2006, from http://www.philipmorrisinternational.com/pmintl/pages/eng/default.asp

Philip Morris USA. (2007). Homepage. Retrieved October 14, 2006, from http://www.philipmorrisusa.com/en/home.asp

Sapienza, P. (2001). Nurturing translocal communication: Russian immigrants on the World Wide Web. *Technical Communication, 48*(4), 435–448.

SDL. (2008). Web globalization. Retrieved June 28, 2008, from http://www.sdl.com/en/solutions/business_need/web_globalization.asp

Sloane, S., & Johnstone, J. (2000). Reading sideways, backwards, and across: Scottish and American literacy practices and weaving the Web. In G. E. Hawisher & C. L. Selfe (Eds.), *Global literacies and the World Wide Web* (pp. 154–186). New York: Routledge.

St.Amant, K. (2002a). Integrating intercultural online learning experiences into the classroom. *Technical Communication Quarterly, 11*(3), 289–315.

St.Amant, K. (2002b). When computers and cultures collide: Rethinking computer-mediated communication according to international and intercultural communication expectations. *Journal of Business and Technical Communication, 16*(2), 196–214.

St.Amant, K. (2005a). Online ethos and intercultural technical communication: How to create credible messages for international online audiences. In C. Lipson & M. Day (Eds.), *Technical communication and the World Wide Web* (pp. 133–165). Mahwah, NJ: Erlbaum.

St.Amant, K. (2005b). A prototype theory approach to international Web site analysis and design. *Technical Communication Quarterly, 14*(1), 73–91.

St.Amant, K. (2007). Online education in an age of globalization: Foundational perspectives and practices for technical communication instructors and trainers. *Technical Communication Quarterly, 16*(1), 13–30.

Ulijn, J. M., & Campbell, C. P. (2001). Technical innovation and global business communication: An introduction. *IEEE Transactions on Professional Communication, 44*(2), 77–82.

Ulijn, J. M., Lincke, A., & Karakaya, Y. (2001). Non-face-to-face international business communication: How is national culture reflected in this medium? *IEEE Transactions on Professional Communication, 44*(2), 126–137.

Webtraduction. (2004). Translation, Web site design, localization. Retrieved June 28, 2008 from http://www.webtraduction.com/index.htm

Yli-Jokipii, H. (2001). The local and the global: An exploration into the Finnish and English Web sites of a Finnish company. *IEEE Transactions on Professional Communication, 44*(2), 104–113.

Yunker, J. (2002). Going global gracefully: Strategies for building the global gateway. Retrieved June 16, 2008, from the Moving WebWord Web site: http://webword.com/moving/yunker.html

Zahedi, F., Van Pelt, W. V., & Song, J. (2001). A conceptual framework for international Web design. *IEEE Transactions on Professional Communication, 44*(2), 83–103.

CHAPTER 2

Making the User the Localization Expert: Employing User-Customization Strategies in Globalizing Online Content

Clinton R. Lanier

CHAPTER OVERVIEW

This chapter outlines the difficulties of delivering online content to a multi-cultured, global audience. Because such an audience comprises a variety of cultural-rhetorical demands, online content must be closely localized to ensure that the information is disseminated accurately and effectively (that is, made appropriate for the culture). However, such close attention to individual cultural-rhetorical preferences has not previously been possible due to the vast amount of resources required. To provide a potential solution to this problem, this chapter discusses user-customizable online technology and suggests that technical communicators can utilize such tools to create information products allowing users to define their own culturally appropriate online experiences.

Past scholarship in Web site design has been a key factor in enabling organizations to most effectively deliver their online content to their site's users. Organizations, as Lin (2002) points out, will have different needs according to many different criteria. The larger an organization, for example, the larger its representative Web site will be. Further, as the organization grows, so do the audiences for the Web site, and these audiences will dictate the design of the site (pp. 37–38). Thinking about these issues is an important first step in the design and creation of any online content.

Designing online content for an international audience—especially for organizations hoping to localize their information—goes beyond past scholarship in Web design and demands that new methods for delivering content be created and applied. This chapter discusses a possible, already–existing solution to the difficulties some organizations may face when localizing online content for an international audience. The first section presents a discussion of past research in Web site design and localization for international audiences. This discussion is not meant to be wholly inclusive, but instead to point out research important to the solution presented here. Following this discussion is a more detailed description of the technology used in the solution I describe. Finally, this chapter ends by demonstrating how to apply this technology to an international online content delivery scenario.

RELATED RESEARCH AND STUDIES

There are thousands upon thousands of Web sites created by and for localized audiences around the world. A first step in identifying ways in which organizations can localize their content is to study how other cultures have gone about communicating with their own members via the Internet. A 1998 study performed by Barber and Badre discussed ways that different cultures (represented in their study by national identities) around the world presented information in their Web sites. The authors found that there were indeed correlations between cultural differences and the Web sites developed and utilized within those cultures. They called the specific items representing these differences "cultural markers," and included in this group items such as the language of the Web site, the colors used on the pages, and the orientation or grouping of the information presented.

Barder and Badre did not spend time speculating why certain cultural markers existed on the Web sites of certain countries/cultural groups and not in others; rather, they set out to demonstrate that the markers simply existed, and they succeeded in doing this. Their study indicates, for example, that information is presented more in a right-to-left textual orientation in the Web sites created in Middle Eastern countries, such as Egypt, Israel, and Lebanon. Hence, the orientation of information presented became a cultural marker.

The existence of cultural markers within Web sites is unsurprising, considering the amount of online information created all over the world. It is a consideration, however, that is of the utmost importance for organizations hoping to deliver content locally in other countries/cultures, especially since, as studies have shown, such cultural differences affect the usability of Web sites (Bourges-Weldegg & Scrivener, 1998; Cyr, Ilsever, Bonanni, & Bowes, 2004; Sun, 2001). One study, for example, demonstrated that culture plays a vital role in how usable a Web site may be (Bourges-Weldegg & Scrivener, 1998). The findings of this research suggest that users misunderstand the meaning of markers from different cultures, and such misunderstandings lead to perceived problems with the site's usability. A similar conclusion was reached in a pilot study conducted by Sun, and findings suggesting that users within culture *preferred* familiar markers were reported by Cyr, Ilsever, Bonanni, and Bowes.

Various researchers (e.g., Hillier, 2003; Sun, 2001) have additionally made the argument for including a variety of cultural markers in multilingual Web sites. Sun studied the Web sites of two large global corporations. One organization had simply translated the text on the site in order to localize it for other nations. However, the other both translated the site and included cultural markers—colors, graphics, and page layout—specific to the nations for which the site was localized.

Sun (2001) found that users preferred the sites localized for their own cultures in terms of both the translation and the inclusion of the cultural markers. Similarly, Hillier (2003) related Web usability factors to cultural markers, and suggests that these markers build a familiar cultural context for users, thus making it easier for them to understand and navigate (and therebyuse) a site with not only translated information but also relevant cultural markers.

These are important points, because they suggest that translation alone is not enough to make a Web site usable by members of another culture. Though language is itself an important cultural marker, language localization is only one element in an array of design considerations that information designers must take into account in order to create online content truly adaptable to other cultures.

Expanding on the above discussions about cultural markers, Marcus and Gould (2000) utilize the findings of researcher Geert Hofstede (1980, 1991) in an attempt to explain why Web sites created in different cultures are different. Specifically, they look for evidence of five different cultural aspects in localized Web sites (that is, Web sites created and utilized within specific countries). These aspects are described in Table 1.

Marcus and Gould's study found that some Web sites display their culture's specific characteristics in the site's design. On a Malaysian university Web site, for example, they point out evidence of high power distance. This characteristic is displayed on the Web site through a concentration on the power structure of the university: the prominent area of the site devoted to the university's seal, graphics of items such as faculty, buildings, and administration. Marcus and Gould compare this Web site to that of a university in the Netherlands—a characteristically low power distance culture. This site, they point out, displays an emphasis on students rather than leaders, and reveals a stronger use of asymmetrical layout (implying a less-structured power hierarchy) among other things. Marcus and Gould present similar contrastive findings for each of the characteristics they focus on, and present a convincing case for their hypothesis.

Focusing specifically on Malaysian Web sites in comparison to U.S. Web sites, Gould, Zakaria, and Shafiz Affendi Mohd (2000) report findings similar to

Table 1. Cultural Aspects Looked For in Web Sites by Marcus and Gould (2000)

Power distance	The extent to which less powerful members expect and accept unequal power distribution within a culture.
Collectivism vs. individualism	The extent to which members of a culture pursue ends for other members of the culture as well as for themselves or primarily for themselves.
Femininity vs. masculinity	The extent to which a culture focuses on traditional gender roles and characteristics.
Uncertainty avoidance	The extent to which members of a culture avoid unknown matters due to feelings of anxiety in the unknown.
Long- vs. short-term time orientation	*Long-term:* Certain cultures have tendencies to practice behaviors aimed at the good over a long period of time (such as patience and hard work), long-term relationships, and practice and practical values. *Short-term:* Other cultures tend to focus on characteristics aimed at creating more immediate results.

those of Marcus and Gould. Drawing again on Hofstede (1980, 1991), as well as Trompenaars (1994), the researchers confirm that cultural characteristics identified in past studies are noticeably present in Web sites created within specific cultures.

A study carried out by Wurtz (2005) involved a cross-cultural analysis of Web sites to determine how far they reflect the high- or low-context predominance of the cultures in which they were created. Using Hall's (1998) theories about *high* and *low context*, as well as many of the variables used by Marcus and Gould (such as the number and type of visuals, the amount and structure of text, and markers such as colors and culturally specific elements), Wurtz examined a number of Web sites and demonstrated the ways in which they reflect particular cultures' preferences and norms for communication.

As defined by Hall, high and low context refer to the amount of explicit information a message will contain. When a message primarily contains explicit information in the form of text, it is said to be low-context communication, because little information is being passed on through other means. However, when the information is derived from elements such as the circumstances of the message or the origin and target of the message, it is said to be high-context communication. Hall suggested that the United States is an example of a culture that predominantly communicates in low-context ways, while many Asian cultures, like Japan or China, would be considered high context.

Wurtz's study focused particularly on many of the visual dimensions of the sites analyzed. These dimensions included the "layout and the use of images, photographs, and animations" (2005, p. 9). Again, the study demonstrated that Web sites can and do reflect cultural variables. Wurtz concluded that the Web sites created within high-context cultures reflected their high-context aspects through the elements they included, such as images reflecting nonverbal communication in subtle ways, for example, through body language, or through the presence of noticeably more images and less text on any given page of a site.

As the studies discussed above and other studies demonstrate, an organization must do more than just translate the language of its content in order to localize; the lessons learned from early research on online content delivery must now be measured against a backdrop of culture. True localization takes into account the cultural aspects of the audience for which it is being localized. But how can organizations truly localize their information? Several studies provide possible methods for understanding how online information must be presented to international audiences.

One study suggests that before organizations create a localized identity, they should first conduct a study of potential users from that culture. Aimed at understanding the context in which users create meaning, and intended to increase a site's potential cultural usability, Bourgess-Waldegg and Scrivener's "meaning in mediated action" (1998, p. 303) approach focuses on users as they attempt to understand potential representations on the site ("help" buttons, for example). Bourgess-Waldegg and Scrivener user tested certain interfaces and interviewed the participants afterward. and they concluded that to ensure cultural usability, designers must use representations (metaphors, images, colors, etc.) specific to the culture and to the meaning they wish to communicate.

Another approach asks information designers to understand and utilize the culturally specific visual representations that create meaning. Drawing on the mainly visual aspects of online information units, St.Amant (2005) uses *prototype theory* to create more culturally relevant Web sites. Prototype theory suggests that each person has a specific ideal representation for objects, against which other objects are measured and categorized. Hence, if a person is shown a visual representation of a dog, that representation is compared to that person's own representation. If enough similarities present themselves, then the individual can agree or assume that the representation is a dog. St.Amant argues that such representations are culturally based, and that understanding and using representations similar to those used within the target culture, make the Web site more usable and culturally acceptable.

It is clear that there is no shortage of instruction for technical communicators and organizations wishing to localize information for specific cultures. However, very little is offered to help them actually deliver the information, and hence localization efforts must still face a number of questions: How many different culturally specific audiences must the information be localized for? How "local" should the information be? What about individuals within a culture who are not necessarily representative of that culture's values (such as people in China, a mainly collectivist culture, who instead are more individually oriented)? In other words, how is the content actually to be delivered?

One option is for the technical communicator to localize every piece of information to every potential culture to which the content is being delivered. But to really localize the information, that is, not only to match the cultural markers discussed above but also to deliver content reflecting deeper cultural aspects like those researched by Hofstede and others, organizations' are faced with a significant problem: The economics of such a task render it almost unfeasible, as there are thousands of languages and cultural identities around the world. Imagine the cost and time involved in creating a Web presence for each of these.

Another option is to do what many organizations currently do: create a Web presence for what is seen as the cultural majority of a region. As Bourgess-Waldegg and Scrivener point out, however, such a localization strategy should not be used to deliver information over the Internet, because cultures are not necessarily "ontologically objective" (1998, p. 289). Instead cultures are "continuously interacting and developing" (p. 289), and therefore a static representation of meaning may become inappropriate after a short while. Also, a cultural identity does not equate to a regional identity. For example, Coca Cola has a Web site localized for China. However, there are many different languages and cultural identities within China.

Hence, such a localization attempt requires designers to generalize about the people within a culture, and such generalizations can lead to the type of usability issues discussed by Bourgess-Waldegg and Scrivener and others, such as confusion caused by representations that do not fit an individual user's schema. Ultimately, because of such generalizations, Web site users may not understand the information in the site, and may not understand what to do with the information, or even how to interact with it.

So at issue, then, is the problem of creating content that is localized for cultural and not regional identities, that is not static, and that does not generalize about members within a culture but is instead culturally appropriate for every user.

One potential solution is to make the user, rather than the technical communicator, the localization expert. In such a scenario, users are given all of the tools needed in order to arrange online information according to their particular—that is, cultural and rhetorical—needs and wants. If given a supply of cultural markers, and then given the ability to arrange, include, or exclude such markers as they see fit, users may create for themselves information units that really work according to their cultural standards.

The remainder of this chapter discusses the feasibility of this solution, utilizing existing technologies to deliver culturally customizable content. I now show that the solution is feasible, by briefly presenting information about current technologies that make it so, and I also present examples of pages using this situation that are in use today.

USER-CUSTOMIZABLE ONLINE CONTENT

The medium for delivering online content does not have to be static. Rather, as pointed out by Sapienza (2004), technology currently available allows information delivered online to be presented dynamically in a number of ways. Certain types of technologies in use allow users to directly modify the content of Web pages and Web sites and create their own, unique experiences. Following a review of user-customizable Web sites, I review two technologies—XML (Extensible Markup Language) and server-side scripting languages—utilized to create such Web sites. However, before beginning this discussion, it is important to point out that the time it takes to publish research such as this normally ensures that the research will lag behind the technological advances or changes that have taken place in the meantime. With this in mind, I have gone to great lengths to ensure that the material presented here is the most current at the time of going to press.

User-Customizable Web Sites

I use the term "*user-customizable*" to denote the ability of a Web site to be changed or modified by the site's users rather than by the designers. According to Perkowitz and Etzioni (2000), customization with regard to a Web site refers to "adapting the site's presentation to the needs of the individual visitor, based on information about those individuals" (p. 153).

Another term often used is "*personalizable,*" as used by Bellas, Fernandez, and Muino (2004) and Manber, Patel, and Robinson (2000). From a cultural perspective, the word customizable is preferable because of the Western belief in universal individuality implied by personalizable. Personalizable denotes the *personal* or *individual*, a concept common in many Western cultures, as pointed out by Hofstede's research. Such an implication may not mesh well with cultures leaning more toward collective beliefs: beliefs that put the group above the individual. As pointed out by Bourgess-Waldegg and Scrivener (1998), representations—such as metaphors, concepts, or labels—rely heavily on culture for the way they are defined by readers; thus information designers should always attempt to choose terms that will not carry with them unintended meanings. The concept of customization, while still implying one can create an individual experience, is perhaps more neutral.

Further, the term "customizable" is meant to suggest a user experience in which the user has control over changes in the online content. This definition is not the same as that of Brusilovsky's "*adaptive*" technology. For Brusilovsky (1996), a technology is adaptive when it builds a model of the user and then applies that model to the user's experience by automatically adapting the content to the said user (as the technology has modeled him/her). When a user begins interacting with an adaptive interface, the interface adjusts itself to what it suspects the user wants or needs, without consent from the user. Such an interface can cause problems when it builds a model of a user that the user does not agree with and then delivers information the user does not want or need. In user-customizable Web sites, the technology does not adapt itself to anything but rather allows itself to be changed by the user drawing upon his/her perceived "self"-model. The user is in complete control of the changes that are made to the site.

Sites with user-customizable features are becoming increasingly common in e-commerce contexts. Amazon.com, for example, includes features that allow users to enter custom preferences like e-mail alerts for new products, or relational searches to see products similar to those they are viewing. Sites similar to Yahoo (that is, Web portals) allow further customization through their personal pages, such as My Yahoo! or My Excite (Perkowitz & Etzioni, 2000).

When utilizing My Yahoo!, for example, users can complete a wide variety of customization tasks, such as selecting settings allowing them to view local news and weather, specific stock quotes and sports scores. Further, users can modify the display of the site, including the colors used and the position of components on the page (Manber et al., 2000; Perkowitz & Etzioni, 2000). When a user visits this site, enters a certain item of information, like a username or password, the user then sees his/her personally customized page, complete with up-to-date information on the components selected (such as stock quotes or local weather).

The variety of tasks users can perform in the above example correspond to the *topology of personalization* identified by Rossi, Schwabe, and Guimarães (2001). These authors' topology includes *link personalization* and *content personalization,* which is further refined into node structure personalization and node content personalization (pp. 276–277).

Link personalization would most closely be associated with the Amazon.com example given above: The page is customized in that it displays links to other pages (showing other products the user may be interested in), but the content remains the same (once on that page, all users see the same product description located in the same place). *Content personalization* can affect either the structure (that is, where things are located, as in the movement of information modules within the My Yahoo! page) or the content (the actual information displayed on the page).

Such technology is not without its problems for designers and users. For designers, especially when designing online, customizable content, it is difficult to truly model the potential actions of a user (Manber et al., 2000). Rossi et al. (2001) point out that the "number of personalization variants seems countless" (p. 275). To accommodate this possibility, Manber et al. suggest that when creating such pages, designers should "design everything for infinite growth" (p. 38).

For users, such potential customization may be intimidating. When faced with too many possibilities, the content could be overwhelming, and the site could be judged too difficult to utilize. If it appears too complicated, users may be unsure about how to utilize the Web site's features and therefore "afraid of breaking something, or getting into a state that cannot be undone" (Manber et al., 2000, p. 37). One possible solution is to force the user into customization options (that is, creating Brusilovsky's adaptive system), but this could also turn users off. Manber et al., the designers of the Yahoo customization elements, give us as an example the automatic association of news with a user due to certain pieces of information that are given, such as the zip code of the user or past Internet searches. The resulting information—they use an example of breaking news about cancer as a result of a past Internet search on cancer—can "confuse the user at best, and at worst, can jeopardize user trust" (Manber et al., 2000, p. 38).

Gradually, while it has not yet been widely used for localizing information, user-driven customization of online content is becoming a very pervasive strategy, especially for e-commerce businesses. The next section discusses two particular technologies utilized in user-customizable settings—XML and server-side scripting. The choice of these two technologies is not intended to limit discussion of user-customization technologies to these specific types, but rather to give readers a better idea of how technologies work generally in customizing online content.

XML and Server-Side Scripting

Extensible Markup Language (XML) and server-side scripting are both technologies utilized extensively when delivering dynamic online content. The technology known as server-side scripting fulfills a user's request by running a script directly on a server to dynamically generate a Web page. This is different from older methods of delivering online content, which relied on information written in HTML to be interpreted by the Web browser to which the content was being sent. In such a case, the content could only be displayed in one way: the way it was defined in the HTML code. If any modification to the page took place, it was accomplished only through the HTML code (for example, a JavaScript program running within the page). However, by using server-side scripting, content can be displayed a number of ways depending upon a number of variables. One example is a site that asks for user input in the form of a series of questions. As the user answers a question, the page changes depending on the answer the user gives. The page thereby dynamically and automatically funnels the user to an end result that is user specific.

Such applications—like the open source server-side scripting language, PHP—have become more advanced since they were introduced in the mid-1990s. Users can now customize pages or sites, even to the point of moving portions of a Web page or including and excluding the information presented. Further, by using server-side scripting technologies, Web pages and sites can "remember" users and the customization preferences of those users. This memory knowledge is stored on the server and is retrieved when a user revisits a site. Users will normally give the server some sort of tracking data, like a username and/or a password—items that Bellas, Fernandez, and Muino (2004) refer to as "personal persistent objects" (p. 234).

Extensible Markup Language (XML) is a markup language that allows designers to provide "structure-less content" with the expectation that when called for, the information will be assembled and structured according to what information is called and how. The information is created in small modules that are "tagged" in such a way as to make them easily stored and recalled. This is similar to HTML or SGML (on which XML is actually based). For example, all customer names could be tagged <custname> and placed in a repository. When someone wants to see all of the customers, a search can be run against the repository for any portions of data matching the tag <custname>. The repository is searched and the data sifted through until all information matching the query guidelines (that is, with the tag <custname>) is returned.

The primary difference between XML and the other two markup languages mentioned above lies in the ability to define XML elements (that is, by creating tags like <custname>). In SGML and HTML, each tag is predetermined by the rules of the language. But because XML was specifically created to allow more flexibility, information designers can create any number of element definitions. This ability allows the data to be sorted, configured, searched, and structured in just about any way imaginable.

To an information designer, this means that the content being delivered does not have to be structured when encoding the online delivery module (that is, when creating the Web page, for example), but can be structured at the time it is retrieved by the online decoder (such as a Web browser).

For a user, this means that the content being viewed can be changed at any time, simply by restructuring it within the Web browser. A user viewing her financial records on her bank's Web site, for example, is able to restructure the table she is viewing to see only transactions between certain dates, or to view only transactions involving more than a specific dollar amount. All of this is done without leaving the page she is viewing, and it is done by organizing the data according to her own parameters. Prior to the introduction of XML, the data either had to be statically displayed or was called from databases residing on servers and viewed in clunky pull-down menus with limited parameters (limited because they were chosen by the designers, not the user). Thus, prior to XML, the designers of a Web site had to anticipate the various ways in which users would want to access data, and the users were limited by the designers' choices.

Many technical communicators are already familiar with the publishing model called single sourcing. In this model, documents are not created linearly, as they have traditionally been created. Instead, information is created and arranged in a modular fashion, and then when a document is created, the modules are assembled in a specific order or arrangement (for more information on single sourcing, see Kramer, 2003; Rockley, 2001). The markup languages SGML and XML allow even words to be part of this modular system. In essence, the solution presented here suggests that technical communicators should approach cultural factors as they would structured content, by providing the user with tools—that is, the technologies discussed in this section—and the ability to provide his or her own *cultural* structure for the information. The following section discusses user-customizable technology in use and shows that it is already being used to deliver culturally customizable online content.

Joining the Two: Using User-Customizable Web Technology to Deliver Information Internationally

User-customizable technology is already being used to deliver content around the world that is structured to be culturally specific. Yahoo has created nation-specific Web portals (which they call Yahoo International) representing countries around the world. Yahoo users can choose to create a My Yahoo! customized Web portal from within these nation-specific Yahoo International sites.

Figures 1 and 2 display the My Yahoo! Web portals for different countries (India and Japan) without any modifications. Any user who selects "My Yahoo!"

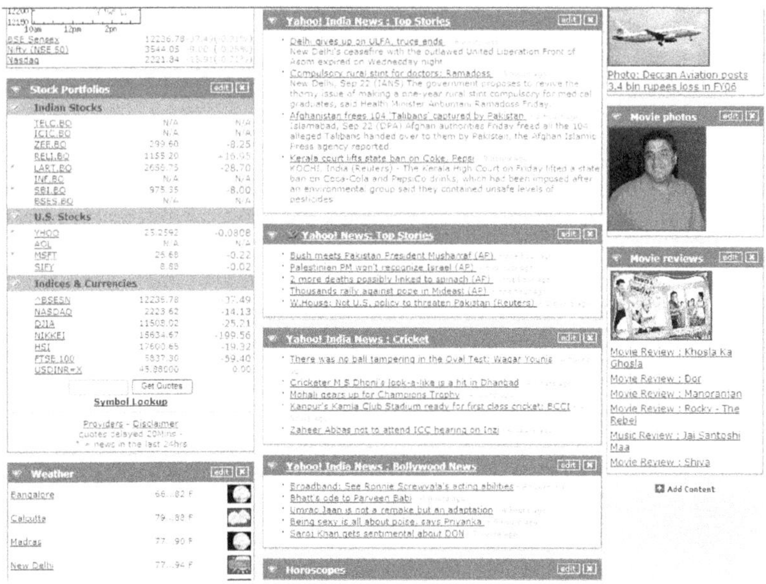

Figure 1. My Yahoo! Automatically generated portal for India.

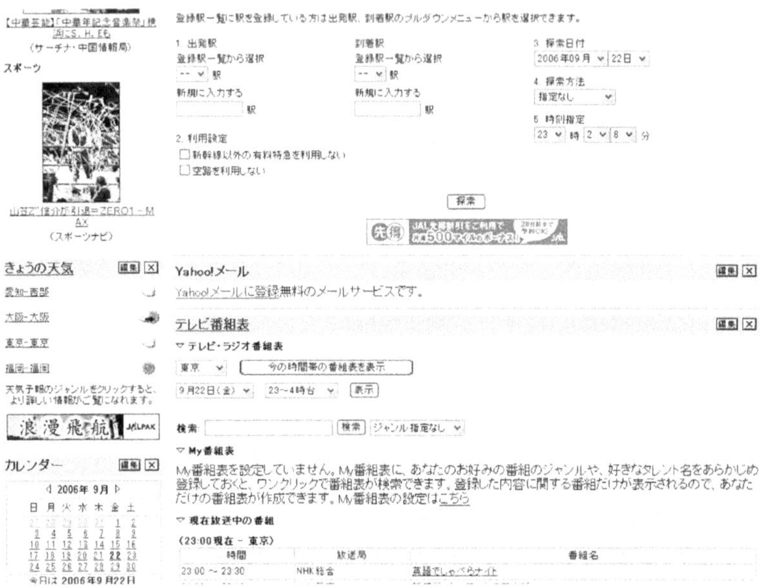

Figure 2. My Yahoo! Automatically generated portal for Japan.

from within "Yahoo! International" portal would see similar pages. As can be seen in both, certain cultural markers are already displayed. The page shown in Figure 1 has automatically generated stock quotes specific to the country represented by the Web portal—in this case "Indian Stocks." In Figure 2, the Japanese portal is displayed in Japanese Kanji script. And in both portals, modules containing localized weather information (or at least weather information specific to the country) and localized headlines are displayed.

The customization features on these pages are achieved through a combination of RSS feeds and server-side scripting. The RSS feeds supply the information on the page, while the server-side scripting allows each element on the page to be dynamically altered in some way (for example, by moving the elements around, changing what elements are displayed, or changing color settings).

Technically, users can customize their My Yahoo! portal to include any cultural marker that Yahoo—or the technology—allows. Should a user prefer less text and more graphics, indicative of a cultural preference for high-context communication (Hermeking, 2005; Würtz, 2005), the user can set the preferences of the My Yahoo! portal in such a way that graphics are displayed with less associated textual information. Similarly, users with a collectivist cultural preference (i.e., having more regard for the collective well-being and placing less emphasis on the individual) can view information about localized sports teams, companies, or organizations, or national and local governments—information that affects the "collective." Users with a preference for individualism can include personal stock portfolios, user-relevant news and weather, and any other information that specifically impacts the user as an individual.

In none of these situations would the geographic settings matter. Users could choose to make My Yahoo! Japan their Web portal, or My Yahoo! India their Web portal. Nor would their own preferences *have* to follow the predominant preferences of their own culture.

And this is perhaps one of the most critical points. Such customization allows the user to decide how information should be presented. As stated earlier, this option eliminates the chance of stereotyping or generalizing about everyone within a country. True, people within a culture will display a strong predominance for certain preferences; however, they will not all feel as strongly toward these preferences, and as individual users they may display these preferences in different ways. Allowing users to customize the display of information—link and node personalization—as well as allowing them to customize the information itself—content personalization—is important to creating online content that is usable for more users.

What is missing from the current delivery of user-customizable online content is the widespread inclusion of *content* personalization. In the case of the My Yahoo! portals, users *can* customize the delivery of news headlines and stories by picking and choosing the RSS feeds included on the page, but they *cannot* customize the way the stories are written. This is an important issue when considering that the way information is structured within cultures—the rhetorical strategies used—could be a major cultural difference between users.

Technical communicators need to understand not only how users may adjust what I will call the "topical" cultural markers of a Web page—items like the colors, text orientation, language, and images, that are in the top or visual layer of a Web page—but also the more complicated markers: the textual content of the page. As discussed by Zahedi, Van Pelt, and Song (2000), cultural factors such as high-power distance, time orientation, and masculinity and femininity can easily be related textually. Zahedi et al. present as an example of masculinity—associated with "qualities of success, assertiveness and strength" (p. 97)—the following bit of text from one of the Web pages they studied:

> Be all that you can be. Become part of the best team in the world. Join the US Army or Army Reserve and we'll guarantee great training and skills to last a lifetime. Learn more and see if you're ready. (p. 97)

Naturally, individual users who share such a cultural view will be attracted to such information.

Barry Thatcher (1999) has also done much to investigate how cultural values influence the rhetorical strategies—how someone reads, understands, and interacts with information—of people within other cultures. His case study of the construction of two documents in Ecuador suggests that people's culture could significantly affect the way they structure and understand written information. Among many other things, Thatcher found the following:

- Certain audiences may prefer *written* information to be composed as *oral* information is presented. In some cases this may mean that information is presented not in the logical, sequentially structured information pattern of information of the United States but in a narrative format that more closely resembles the moral-based stories of some cultures.
- Though most South American cultures are considered high-context, they are also considered to be high–uncertainty avoidance cultures (that is, they prefer information in contexts that are familiar, and struggle in unfamiliar rhetorical situations). In Thatcher's case study, this finding was represented by the participants' struggles to apply context to abstract rules. They wanted more concrete examples of the rules in situations they would face.
- Within the documents that were created in the course of his study, Thatcher noted that participants had a preference for "accumulative rather than analytic and hierarchical communication patterns" (p. 184). Instead of dividing a problem or guideline into different categories to be examined, the South American participants preferred more repetition and addition of information. Hence, each section of the document repeated prior information and introduced and clarified new information.

Though, as stated above, users can customize the topical cultural aspects that Barber and Badre and others have discussed, the culturally specific rhetorical strategies noted by Thatcher, Zahedi et al., Hofstede, and Trompenaars may be even more important. Cultural preferences for color, directionality, and location of information may help to make a Web site more usable and may also enhance the "e-loyalty" of a Web site (as discussed by Cyr, Ilsever, Bonanni, & Bowes, 2004), but users may need even more customization abilities to allow for the their' preferred rhetorical strategies.

The following section utilizes Thatcher's findings to present an example of such a solution in action. Drawing from Thatcher's article on technical communication in a South American context, this section presents a scenario in which hypothetical content must be delivered to an audience similar to the one described by Thatcher.

UTILIZING THATCHER'S FINDINGS TO DELIVER USER-CUSTOMIZABLE ONLINE CONTENT

It would be valuable to see how an organization could practically deliver user-customizable online content that considers both the topical cultural aspects discussed in the introduction and the deeper cultural-rhetorical aspects introduced in the previous section. In order to do this, I present a hypothetical situation drawn from the case Thatcher studied, with the intention of showing how the organization in that case could effectively deliver the information to each user while considering the culturally specific learning styles and preferences.

In Thatcher's case study, an organization headquartered in the United States wanted to standardize its global accounting customs. To do this, the organization created a set of standard accounting practices to be taught to and adopted by accountants in five countries in the South American region: Colombia, Ecuador, Venezuela, Peru, and Bolivia. Thatcher's study—and subsequent recommendations—revolved around the problems encountered when the training and rhetorical styles used by the organization's officials differed from the learning and rhetorical styles of the representatives of the five countries.

The organization's officials were trained in the United States and thereby naturally utilized the cultural-rhetorical patterns they were used to. However, with the exception of two, the organization's South American accountants had different cultural-rhetorical patterns. The two exceptions, Thatcher explains, had patterns more closely resembling those of the U.S. personnel: linearly structured procedures featuring abstract guidelines. These two stood out because while the other South American accountants struggled with the lessons and guidelines being imposed, Guillermo and Jaime (the two exceptions) were able to grasp the concepts in the manner in which they were taught. These two, as opposed to the other accountants, had a great deal of exposure to U.S. rhetorical patterns, and so found the rhetorical patterns of the organization's officials more familiar and easier to understand.

This example makes clear the importance of not making generalizations when localizing information. Had the organization decided to create online tutorials to train the accountants, they might have used many of the cultural markers pointed out by Barber and Badre (1998). They might have also used some of the deeper rhetorical strategies that Thatcher points out are important in helping the South American accountants understand the new policies. However, in such a case, a generalization of this kind—cultural markers and cultural-rhetorical patterns of information intended to help the South American accountants—would have left out Guillermo and Jaime.

This example actually warrants the type of solution I am arguing for: online content delivered via user-customizable technology. Let's suppose that the organization Thatcher had studied *had* decided to create an online tutorial to teach the South American accountants the new standard practices being implemented. Users of the tutorial could start off by utilizing the type of *customizable modularity*—or node structure personalization—seen in Web portals such as My Yahoo! and My Excite.

Such changes are easily facilitated by utilizing the server-side scripting technologies discussed earlier. The accountants would be able to customize topical cultural markers like the tutorial's interface by moving chunks of information so that spatially the interface was in line with cultural preferences. Specifically, information could be presented on the left or right side of the screen, and more or fewer graphics or textual elements could be presented according to the user's desire. The accountants could also customize markers like the colors of the interface and the language in which information in the tutorial is presented. All of these cultural markers are important elements in "culturability," according to Barber and Badre (1998).

Further, utilizing link personalization and node content personalization methods, facilitated by both server-side scripting and XML technologies, users could also effectively customize the ways in which subjects and information are presented. Instead of moving through a linearly structured hierarchy of information, the content could be customized by the user so that information is repetitive and accumulative. Each step, if the user so chose, could be presented as repeating what the prior step listed, and build on that information by presenting new items. This presentation of information would closely resemble the more oral patterns pointed out by Thatcher in the examples he cited.

Finally, users could customize the content of the tutorial further by choosing settings to allow for more or fewer examples of the guidelines being put into practice. Thatcher argues that such contextualization helps members of high-context cultures orient themselves to new information. The contextualization that was needed was frustrating to the organization's U.S. personnel, because while they were attempting to teach the abstract principles of the guideline being presented, the accountants wanted concrete examples of the guideline in practice. The need to ground an abstract principle could easily be accommodated by the hypothetical online tutorial being discussed here.

Figure 3 presents a visual representation of the processing sequence for customizing tutorials with culturally specific rhetorical strategies. Guillermo and Jaime are represented by user A, and the rest of the accountants by user B. Each group would begin the online tutorial by looking at the default interface. As the tutorial began, they would select from choices they had been given about how they wanted the information presented. The choices would result in calls to the server for specified modules populated by information structured in particular ways that match culturally specific rhetorical structures (real examples, narrative-based instructions, linearly-structured procedures, etc.). The result would be a uniquely structured interface for each group (or individual) that reflects the choices that had been made by the user and that would enable him/her to best understand the information.

Of course it would be a mistake to assume that users would be aware of their own culturally specific rhetorical and communication patterns. Researchers are still in the process of understanding how different cultures communicate, so we cannot expect users to know what they want or need out of an informational unit to enhance their understanding. However, the method in this scenario, the

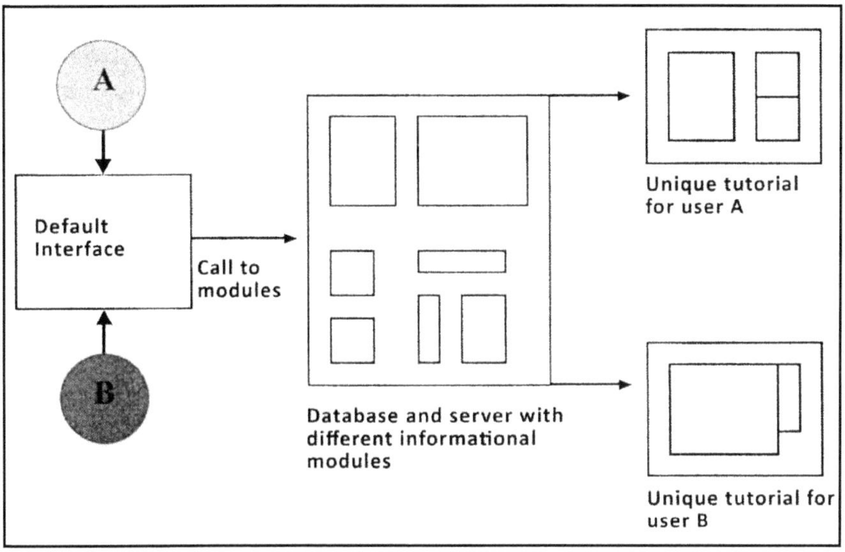

Figure 3. Process by which different users customize their own interface with culturally specific rhetorically structured information.

method I am suggesting here, is one in which users have complete control over the customization of the interface.

As mentioned previously, other technologies and methods—like adaptive interfaces—have also been created, but these technologies depend on either generalizing the user or interfacing with users who already know what they want. In the small case study Thatcher discussed, both techniques would fail. An interface that generalizes would assume that because users are from a certain geographical region, they would all want information structured in a certain way. The method advocated here suggests that users would come to find the rhetorical structure they prefer by selecting that rhetorical structure over other structures. After all, in Thatcher's case study (and also as suggested by the many studies cited above), the users who did not understand the American rhetorical structures (linearly structured procedures with no grounding) knew what they wanted: more real examples and information explained to them in an oral pattern.

XML in combination with a single-sourcing approach to delivering information could be utilized to parse and deliver each of the culturally/rhetorically important sets of textual information. A technical communicator could tag passages—such as lengthy examples, or orally structured explanations—so that the text would be included or excluded according to the users' preferences. Hence, multiple explanations could be written and saved, and then delivered to the appropriate audience at the specified time within the tutorial.

Each of these scenarios-within-the-scenario could be saved as the user's preference by using personal persistent objects—such as usernames or passwords. Further, and perhaps most importantly, they would be *options* that the user chooses. The organization could therefore distribute the online tutorial and all of the accountants could utilize it, including Guillermo and Jaime, because while it contains the recursive rhetorical patterns prominent in contexts found in South America, and an extensive collection of concrete examples to help orient users to unfamiliar information, these items are displayed only at a user's request. Thus, if Guillermo and Jaime prefer the rhetorical patterns of the U.S. organization, their customization preferences would be set to accommodate them, and information would then be displayed sequentially, hierarchically, and more abstractly.

CONCLUSION: EXAMINING CULTURE AND SINGLE SOURCING AND REEXAMINING THE USER'S ROLE IN INFORMATION DELIVERY

A solution like the one explored in the above scenario is certainly technologically possible via server-side scripting and XML. Additionally, the solution also resolves many of the problems identified by Bourgess-Waldegg and Scrivener (1998). Culturally relevant information would not be static, for example, because users within a culture may adjust the information delivered to them as they see fit. Likewise, organizations could feel more confident that they are not leaving out certain users due to the cultural generalizations that information designers were forced to make when targeting the majority of people in a region.

With this solution's advantages, however, also come some very important complications. At what level should organizations attempt these customization techniques? There would be a valid argument against customizing every commercial Web site in this manner, as it might not be needed. However, commercial Web sites are not the only information units distributed online. A great deal of online content is disseminated to various groups around the world, where approach suggested above would be warranted. Examples would include health-related information and training, risk management information and announcements, and general policy and procedural information for the operation of global commercial entities. In each case, different audiences—audiences that are culturally different—would depend on correctly understanding the information. Determining what information should be distributed in this manner could come down to costs, as it would require more effort than simply translation to achieve such distribution.

Another difficulty lies in actually organizing the rhetorical patterns to allow for the many different cultures that potential users may represent. Related to this is the need to determine exactly what part the user should play in this process.

In the example used above, only two cultural identities (those of the United States and South America) were discussed, but imagine the distribution of a tutorial to China, Japan, India, and Germany. Each global region contains different cultural identities with differing world (that is, cultural) views. It would be the responsibility of the technical communicators to create different versions of the same information in a variety of languages, as well as to create and include the various needed cultural markers, in order to compensate for the many different ways in which users may want to view the information.

Additionally, difficulties would lie in actually ensuring that the information is localized. In the scenario presented above, examples were integrated (according to the users' customization parameters) to compensate for the users' desire for "grounding" the abstract information presented. But to make these examples effective, cultural experts would be needed to create *culturally significant* examples. Also, the display of information would have to match the dominant patterns in each culture: narrative structures differ from region to region, and this rhetorical structure would have to be allowed for.

Problems also reside in giving the user too much choice. As mentioned before, some users may be overwhelmed a wide variety of options. Manber et al. (2004) of Yahoo state that there are users who want to push the limits of customization and others who do only the bare minimum. Such differences may lie in cultural preference for uncertainty avoidance. While some cultures display a preference for exploring unknown features and solutions, others want what is proven and has already been established. In the latter case, it must be determined if a user-customizable solution to delivering online content is the right method.

Further, there is little information on how user-customization affects the usability of interfaces, and whether or not users will like such an approach. But even with little feedback on the solution's practicality, however, the use of user-customizable Web interfaces been briefly suggested by others. Shneiderman and Hochheiser (2001), for example, suggest that user-customizable interface features could be used to compensate for the global diversity of user preferences by enabling users to customize an online experience. Such customization features, they suggest, should be made *universally usable*, that is, able to be used by anyone entering the site. Interestingly, Shneiderman and Hochheiser's article focused on increasing the usability of Web sites as they are distributed globally, implying that the customization approach elevates the usability of the Web site.

User-customizable interfaces have also been suggested for technologies that support synchronous, distributed group collaboration. Marsic and Dorohonceanu (2003) suggest that user-customization be utilized for an interface allowing distributed groups to work together through a shared space. This solution, they suggest, allows users to avoid the imposition of a single interface that does not quite match anyone's need because it attempts to encompass everyone's needs. Instead, users can create an interface that more closely resembles their ideal experience, hence making the interface more usable.

Echoing a call for user-customizable interfaces to increase usability, Perlman (2002) suggests that Web designers build templates from which users can pick and choose elements to create their interface. Again, his suggestion is that an interface that can be tailored to specific users is more usable than a generic interface that tries to match every user's needs.

Though at the outset it may seem obvious—as it does to researchers like those above—that this solution will increase usability, studies *do* need to be completed to ensure that customizable features do not decrease the usability of an interface. Such studies must take into account what the user wants to achieve or accomplish when using the interface. Hence, in the solution presented here, different types of online content must all be studied. For example, users of online tutorials will have different expectations of what they want to accomplish than users of a commercial Web site. User-customizable features in the delivery of online tutorials must be studied in their own right, to determine whether such a technology is usable in a culturally diverse environment. Studies must take into account a culturally diverse group of participants. Usability studies that focus on participants from only one cultural region or identity produce results that may not necessarily apply to different regions or identities.

Should the proposed solution prove effective in the face of these many questions, technical communicators will be responsible for many new aspects of the design and delivery of online information. Giammona's (2004) article forecasting the future of technical communication covers—out of many—three areas that I see as relevant in this solution, areas that Giammona's participants suggest will be important to future technical communicators. Thus, there are three tasks that technical communicators must undertake to make the proposed solution feasible.

The first of these involves culture. Joann Hackos is quoted in Giammona's article as saying that as a profession, technical communication must begin "extending our customer analysis reach outside of the U.S." (p. 355). So although keeping the audience in focus while designing information is already a vital task of technical communicators, it is an even more important task in this solution. While the audience can determine how the information is designed and graphically delivered, the information elements that make up the larger document the reader sees need to be created with cultural differences in mind. Hence, technical communicators will—even more than before—need to be aware of possible cultural differences and the ways in which those differences manifest themselves rhetorically.

This is achieved only through a thorough understanding of how information is rhetorically structured for different audiences. While this sounds like an immense effort, even compensating for a limited number of rhetorical patterns—as in the example given above—would be better than strict translation alone. Further, applying the concepts of single-sourcing (creating one set of files for multiple delivery types), a concept widely practiced in technical communication, could easily be applied to this solution.

The second task is to focus on experience as much as information. Giammona (2004) discusses the need for technical communicators to move "into the realm of user experience design" (p. 353). The solution described here requires technical communicators to think about the user's experience as a starting point. As discussed above, there are still many questions about usability when it comes to user-customizable interfaces, and such questions will be left to technical communicators to answer through user-testing and feedback. Hence, another avenue for future research—aside from better understanding different rhetorical patterns—is found in testing the usability of such interfaces with culturally diverse audiences.

The third task requires technical communicators actually to learn to design appropriate interfaces. Giammona's participants advocate the technical communicators taking a bigger role in the decision-making process on their products. The solution suggested here necessitates that technical communicators become the experts and thereby the primary decision makers on the design and delivery of online information as a product. It means that technical communicators must learn and apply the technology discussed here, and that they must analyze a population of culturally diverse users and help them develop for themselves rhetorically and culturally relevant online experiences.

In conclusion, delivering user-customizable online content is an already existing solution to the very real problem of making information available to a global audience. While I agree with Sapienza's argument that information designers should not place the burden of content structure on the user, this solution, even with its problems, is still better than generalizing or stereotyping individuals within a culture (which is thought by some to be a better option than the alternative: completely excluding some audiences). As user-customizable technologies become even more interactive, information designers will be able to create a variety of ways to allow users to create their own online experience and better serve a truly global audience.

REFERENCES

Barber, W., & Badre, A. (1998). Culturability: The merging of culture and usability. Paper presented at HFWeb '98. Retrieved October, 2004, from http://www.research.att.com/conf/hfweb/

Bellas, F., Fernandez, D., & Muino, A. (2004). A flexible framework for engineering "my" portals. In *Proceedings of the 13th International Conference on the World Wide Web* (pp. 234–243). New York: ACM.

Bourges-Waldegg, P., & Scrivener, S. A. R. (1998). Meaning, the central issue in cross-cultural HCI design. *Interacting with Computers, 9*, 287–309.

Brusilovsky, P. (1996). Methods and techniques of adaptive hypermedia. *User Modeling and User-Adapted Interaction, 6*, 87–129.

Cyr, D., Ilsever, J., Bonanni, C., & Bowes, J. (2004). Web site design and culture: An empirical investigation. Paper presented at the International Workshop for the Internationalization of Products and Systems, Vancouver, BC, Canada.

Giammona, B. (2004). The future of technical communication: How innovation, technology, information management, and other forces are shaping the future of the profession. *Technical Communication, 51,* 349-366.

Gould, E. W., Zakaria, N., & Shafiz Affendi Mohd, Y. (2000). Applying culture to Web site design: A comparison of Malaysian and US Web sites. In *Proceedings of IEEE Professional Communication Society International Professional Communication Conference and Proceedings of the 18th Annual ACM International Conference on Computer Documentation* (pp. 161–171). New York: ACM.

Hall, E. T. (1998). The power of hidden differences. In M. J. Bennet (Ed.), *Basic Concepts of Intercultural Communication* (pp. 53-67). Yarmouth, ME: Intercultural Press.

Hermeking, M. (2005). Culture and Internet consumption: Contributions from cross-cultural marketing and advertising research. *Journal of Computer-Mediated Communication, 11*(1). Retrieved from http://jcmc.indiana.edu/vol11/issue1/hermeking.html

Hillier, M. (2003). The role of cultural context in multilingual Web site usability. *Electronic Commerce Research and Applications, 2*, 2–14.

Hofstede, G. (1980). *Culture's consequences: International differences in work-related values.* Thousand Oaks, CA: Sage.

Hofstede, G. (1991). *Culture and organizations: Software of the mind.* London: McGraw-Hill.

Kramer, R. (2003). Single source in practice: IBM's SGML toolset and the writer as technologist, problem solver, and editor. *Technical Communication, 50*, 328–334.

Lin, C. (2002). Organizational size, multiple audiences, and Web site design. *Technical Communication, 49,* 36–44.

Manber, U., Patel, A., & Robinson, J. (2000). Experience with personalization on Yahoo! *Communications of the ACM, 43,* 35–39.

Marcus, A., & Gould, E. W. (2000). Cultural dimensions and global Web user interface design. *Interactions, 7,* 33–46.

Marsic, I., & Dorohonceanu, B. (2003). Flexible user interfaces for group collaboration. *International Journal of Human-Computer Interaction, 15*, 337–360.

Perkowitz, M., & Etzioni, O. (2000). Adaptive Web sites. *Communications of the ACM, 43,* 152–158.

Perlman, G. (2002). Achieving universal usability by designing for change. *IEEE Internet Computing, 6,* 46–55.

Rockley, A. (2001). The impact of single sourcing and technology. *Technical Communication, 48,* 189–193.

Rossi, G., Schwabe, D., & Guimarães, R. M. (2001). Designing personalized Web applications. In *Proceedings of the Tenth International Conference on the World Wide Web (WWW10)* (pp. 275–284). New York: ACM.

Sapienza, P. (2004). Usability, structured content, and single sourcing with XML. *Technical Communication, 51*, 399–408.

Shneiderman, B., & Hochheiser, H. (2001). Universal usability as a stimulus to advanced interface design. *Behaviour and Information Technology, 20,* 367–376.

St.Amant, K. (2005). A prototype theory approach to international Web site analysis and design. *Technical Communication Quarterly, 14,* 73–91.

Sun, H. (2001). Building a culturally competent corporate Web site: An exploratory study of cultural markers in multilingual Web design. In *Annual ACM Conference on Systems Documentation* (pp. 95–102). New York: ACM.

Thatcher, B. (1999). Cultural and rhetorical adaptations for a South-American audience. *Technical Communication, 46,* 177–195.

Trompennars, F. (1994). *Riding the waves of culture: Understanding diversity in global business.* New York: Professional Publishing.

Würtz, E. (2005). A cross-cultural analysis of Web sites from high-context cultures and low-context cultures. *Journal of Computer-Mediated Communication, 11*(1). Retrieved from http://jcmc.indiana.edu/vol11/issue1/wuertz.html

Zahedi, F. M., Van Pelt, W. V., & Song, J. (2000). A conceptual framework for international Web design. *IEEE Transactions in Professional Communication, 44,* 83–103.

CHAPTER 3

Optimizing International Information Systems

Matthew McCool

CHAPTER OVERVIEW

Optimizing international information systems depends on a number of rhetorical aspects for engaging a target audience. In the past, development models have emphasized currency, time, and translation in internationalizing online systems. While time and translation are necessary features for international information systems, they are comparatively superficial features. For optimization to occur, system developers need to know much more about their target audience, and this means developing an intercultural theory of mind. This chapter addresses (1) a computational theory of mind, whereby the mental life is likened to information processing and computation; (2) emergent cultural properties, which are the deepest dimensions of culture; and (3) international optimization through revised taxonomies and search engines.

The greatest challenges facing both online developers and internationalization personnel is not a matter of translation, currency, and time (Aykin, 2005; Aykin & Milewski, 2005; Marcus, 2005; Morville, 2005). Rather, the greatest challenges are rooted in a need to adapt online content to the unique communication demands of the target user. From an international perspective, these challenges are best addressed through a three-part process. First, one needs to have a firm understanding of how the mind works, and the best general-purpose model is the computational theory of mind. Second, the computational theory of mind must be divided into two parts, conscious and unconscious decision-making. This merges the computational theory of mind with the deepest dimensions of culture, thereby developing an intercultural theory of mind. Third, this intercultural theory serves as a guide for modifying taxonomies and retrieval algorithms. Thus, the process of internationalizing information systems should assess a number of cognitive and intercultural factors that have previously been overlooked or dismissed (Morville, 2005).

The reason why an intercultural theory of mind is necessary for optimizing international information systems is based on the assumption that culture informs language and linguistic classification. In turn, linguistic classification influences online classification, which is also reflected in search engine adaptations. The overarching goal is based on the belief that culture has an important influence in terms of the way people view the world, that culture is connected to language, and that the varieties of language use affect labels, categories, and search terms.

The goal of this chapter is to present a revised framework for optimizing international information systems. A computational theory of mind is useful for understanding culture. And culture is useful for devising effective strategies for internationalizing information systems. This approach involves:

- *Computational theory of mind:* This is based on the assumption that the mind is like a computer with specialized modules designed to perform specific tasks. The mind is a massively parallel system of computation that makes two different kinds of decisions, analytical and unconscious. Analytical thought is slow and logical, while the adaptive unconscious is rapid and intuitive.
- *Culture:* Culture resides in the adaptive unconscious. People do not need to use logic to adhere to cultural norms. Rather, the deepest dimensions of culture are deep and enduring, making them automatic.
- *Optimizing international information systems:* Users of online systems rely on a combination of analytical and unconscious decision-making. The adaptive unconscious is typically overlooked in user experience studies, which is an important omission because culture is installed in this conceptual area of the mind.

The chapter begins with a description of the latest general-purpose understanding of the mind, as seen through computation and specialization. Next, the adaptive unconscious, the conceptual and most ancient aspect of the mind, which serves as a kind of autopilot, is considered. The adaptive unconscious is important because it houses the deepest dimensions of culture. While many cultural values are useful for international optimization, this chapter examines the two most important, social relationships and uncertainty. The final section of the chapter offers preliminary considerations for achieving an international information system, organizational taxonomies, and retrieval algorithms (cf.

Buckland, 1991). Thus, the objective is to describe rather than prescribe the optimization of international information systems.

THE COMPUTATIONAL THEORY OF MIND

The computational theory of mind assumes that mental life is best understood through the concept of computation. Data is received from the environment, the mind processes this information, and then the body produces some kind of response. The computational theory of mind is useful because it accounts for the concept of specialization. Examples of specialization include an intuitive understanding of logic, number, friends, foes, family, food, sex, shelter, and safety, to name a few (Pinker, 2005b). Culture accounts for another type of mental specialization, and it applies to information systems. Thus, forming an optimal international information system requires a basic understanding of how the mind works and, most important, how cultural software affects mental hardware (Hofstede, 2005).

One way of thinking about the mind is through the analogy of an information processing unit (Forbes, 2005; Pinker, 2005b; Seife, 2006). This means that the mind partly follows tacit, naturalized, and nonconscious thought patterns. While the mind makes reasoned analytical decisions, it is also capable of responding in a rapid manner. If attacked by a lion, for instance, one does not need to exhaustively scan one's mental database in order to respond. One reacts automatically, and there are good reasons to suspect that the user experience is also characterized by some of these same decision-making patterns (Morville, 2005). A prime example of automatic unconscious thought occurs when a user leaves an online system in a matter of seconds, content that he or she has adequately scanned the search results by looking only at the first page.

The concept of specialization is based on a modular sense of the brain. This means that the mind consists of a series of conceptual modules dedicated to specific tasks. Although everyone needs to be taught finite mathematics, most people have no trouble reasoning that walking over a cliff results in death. Human nature has equipped people with many innate modular and specialized abilities, and one of the most interesting and best studied is the visual system.

The human visual system is a sophisticated module dedicated to the many different tasks associated with visual processing. The visual cortex is the largest module of the human brain, and also one of its oldest. The visual system processes information via the visual cortex, which may be divided into five visual areas (Edelman, 2005). V1 (visual area 1) is the central processing unit of the visual system, which is responsible for assessing the status of inert and kinetic objects. V1 is also implicated in pattern recognition. V2 specializes in the recognition of complex patterns, orientation through time and space, and color. V3 computes vector orientation, including the path, projection, and velocity of things. V4, like V1, processes spatial orientation and color. Thus, analysis of color is processed through V1, V2, and V4. And V5 appears to process complex motion throughout the visual field.

The visual cortex is one kind of specialized module in the brain that works in tandem with other specialized modules such as the hippocampus to coordinate spatial navigation (Abramov, 1997; Davidoff, 1997; Edelman, 2005). As with the visual cortex, the role of specialization in the computational theory of mind depends on a combinatorial interplay between modules. Culture is conceptually similar to visual processing, in that an area of the brain is responsible for retaining the most deeply held values and beliefs of a people. And this means one needs a good sense of the source of culture.

CULTURAL VALUES AND BELIEFS

Determining how to optimize information systems for international audiences requires an understanding of the audience's naturalized assumptions about how the world works. This is the intercultural theory of the mind. Naturalized assumptions reflect a way of understanding the world. Such a lens combines both innate and learned behaviors; we will examine the social or learned aspects of core dimensions.

Deep cultural values frequently go unnoticed by the very groups of people to whom they apply, residing in the conceptual area of the mind known as the adaptive unconscious (Damasio, 2005; Edelman, 2005; Goldberg, 2002; Wilson, 2002). The adaptive unconscious is sometimes difficult to understand, because it seems to be based on an abstract notion of the mind. In actuality, the adaptive unconscious is analogous to a kind of mental autopilot in which one need not exhaust a database of logical rules for making decisions. Tying one's shoes or fastening a seat belt occur without the individual's having to explicitly think about performing these activities. Learning how to serve in tennis may begin with a series of lessons and conscious attention, but it eventually gives way to automatic thinking. A tennis serve just happens. Culture is similar, because initial learning is eventually replaced by naturalized assumptions.

The Adaptive Unconscious

The adaptive unconscious is a conceptual area of the brain known for making automatic decisions (Wilson, 2002). Although studying the adaptive unconscious is a fairly recent phenomenon, especially prominent within the last 10 years, enough has been learned to build a useful heuristic (see Table 1).

Conscious thought is characterized as a single mental system, as in analytical reasoning in the frontal lobes (Goldberg, 2002). Conscious thought is also responsible for the deep analytics associated with verifying data, and is responsible for long-term orientation and contingency planning. The slow and intentional nature of controlled procedures is likely a function of conscious thought (cf. Pinker, 2005b). Conscious thought also develops more slowly than

Table 1. Differences Between the Adaptive Unconscious and Conscious Thought (adapted from Wilson, 2002)

Adaptive unconscious	Conscious thought
• Multiple systems	• Single system
• Online pattern	• Posthumous check and balancer
• Concerned with the present	• Future orientation (long view)
• Automatic (fast, effortless)	• Controlled (slow, intentional)
• Rigid	• Flexible
• Precocious	• Slower to develop
• Sensitive to negative information	• Sensitive to positive information

the adaptive unconscious but is especially sensitive to signals of reward (as found in the amygdala).

The adaptive unconscious is characterized by several important features that are distinct from conscious and calculated thought. First, the adaptive unconscious is responsible for lower order mental functions occurring outside of awareness, as in the human visual system. Nonconscious thought is also automatically driven by the mind's autopilot, the instantaneous response to environmental information. The adaptive unconscious serves as an active database of stereotypes that define how we classify and categorize information, creating habituation of thought. One of the more controversial claims with regard to the adaptive unconscious is that it can harbor, even project, deep-rooted feelings and preferences for particular beliefs and actions—feelings that defy rationality (Davidson, 2001). It is important to understand that the adaptive unconscious provides the groundwork for understanding cultural values and beliefs, such as the sense of self in relation to groups, and the way of coping with uncertainty. The adaptive unconscious is considered part of the limbic system.

In evolutionary terms, the limbic system and its adaptive unconscious are the oldest features of the brain organ in *Homo sapiens* (Damasio, 2005; Edelman, 2005; Goldberg, 2002; Wilson, 2002). By contrast, the most recent evolutionary brain structures are the neocortex and frontal lobes. While it had long been assumed that the neocortex and frontal lobes were solely responsible for rational discourse—consider Descartes' separation of mind and matter—recent inquiries by evolutionary psychologists and neuroscientists cast a rather different perspective on the role of what is commonly thought of as lucid thinking (Damasio, 2005; Pinker, 2005a; Wilson, 2002). In short, recent research supports the claim that much of the cognitive continuum that consists of processing, analyzing, and reacting to environmental information occurs within the ancient limbic system (Damasio, 2005; Edelman, 2005; Wilson, 2002). Understanding the intersection between culture and the adaptive unconscious is critical for internationalizing information systems, because the user experience is often swift and automatic.

Two cultural values found in the adaptive unconscious are especially useful for understanding how one should reconsider an international information system. One is the way culture affects strategies for dealing with uncertainty. The other is one's sense of self and how it relates to groups of people.

Individual and Group Identity

Cultures exhibiting higher degrees of personal freedom tend to emphasize equality, individual achievement, clock rather than social time, action through verbs, and task completion (Hofstede, 2005). The presumption of equality is more of a metaphysical ideal than a real set of behaviors. Even in group-oriented organizational settings, personal achievement is emphasized in individualistic cultures. Individualistic cultures are also associated with clock time rather than human time, and achievement is materialized through actions and verbs. Individualistic cultures are also task oriented, which means the emphasis is on completing specific procedures for material benefit, product, or reward.

Collective cultures emphasize group harmony, group achievement, social time rather than clock time, nouns, social status, and topic orientation (Hofstede, 2005). In group-conscious cultures, the presumption of social hierarchies is expected rather than dismissed. Group achievement is emphasized in collective cultures. Also, writing plays a secondary role in exchanging knowledge, presents few opportunities for revealing truth, tends toward the abstract and external rather than the deep and analytical, and omits details, in order to aid in conformance to existing ideals.

The distinction between individual and group identity reflects the perceptual distance between the self and the social fabric. The distance between the self and the social has profound implications for information systems. Attitudes to uncertainty, like personal and group identity, are developed in accordance to the rules and guidelines of local cultural values and beliefs.

Avoiding the Unknown

Uncertainty avoidance is the degree to which one avoids uncertain or ambiguous situations (Hofstede, 2005). Although uncertainty is a human universal, the extent to which uncertainty is handled is partly a function of one's environment. Examples of environmental factors that predispose people toward higher uncertainty include war, famine, poverty, political instability, and public health problems. Extreme cases of uncertainty lead to unhealthy states. People may appear occupied, unfocused, emotional, and active in places where uncertainty avoidance is strong (Hofstede, 2005). Although uncertainty comes in many varieties, relief from it occurs mainly through technology and science, a reliable legal system, and religion (Hofstede, 2005).

Quantitative methodologies reflect tendencies of low uncertainty, primarily through a willingness to form conjectures, test them, and accept the possibility of their falsity. One of the guiding principles toward knowledge in low uncertainty avoidance cultures are ideals of quantification, various forms of statistical analysis—including Bayesian reasoning (Todd & Gigerenzer, 2000)—a general interest in crunching numbers, and decision-making based on fewer sources of analytical evidence (cf. Goldman et al., 1996). Conventional examples of low uncertainty avoidance cultures are found among garden variety top 10 lists (biggest, best, most, fastest, efficient, and so forth). Despite its avoidance in the humanities, quantitative research continues to occupy a high status among research methodologies in low uncertainty avoidance cultures.

Functional uses of statistical quantification may be found on Web sites and in similar information systems such as Amazon.com, which includes a range of statistical information on its products. Examples include a concordance of the 100 most frequently used words; text statistics such as readability (for standard, Flesch, and Flesch-Kincaid indexes); complexity (complex words, syllables per word); number of characters, words, and sentences; and fun stats (words per dollar, words per ounce). Amazon.com also offers a list of statistically improbable phrases (SIPs) and capitalized phrases (CAPs).

While the data are interesting, their usefulness remains to be seen. Even the desire to quantify texts using word frequencies, readability, complexity, and number of characters indicates a strong inclination toward numerical evaluation. While qualitative analyses have blossomed in the United States, especially since the rise of ethnographic inquiry among anthropologists, quantitative methods continue to be prized by private and public organizations alike (Hofstede, 2005). Amazon.com's interest in statistical probability further hints at low degrees of uncertainty avoidance due to the fact that numerical evaluation is considered a powerful tool for testing hypotheses (Quine, 1998). Although Amazon.com presents an interesting example of U.S. tendencies toward quantification, the more interesting product of uncertainty is its relationship with empirical testing.

Theory embodies an area of thought that dismisses or avoids immediate empirical testing. Abstruse philosophical doctrines such as Nietzsche's *Übermensch*, Marxist doctrine, or even the unleashing of Hobbes' Leviathan under particular social dynamics all reflect big ideas that began in high uncertainty avoidance cultures (Hofstede, 2005). Again, a willingness to form postulates, theorems, and hypotheses amenable to empirical testing reflects a fairly low degree of uncertainty.

OPTIMIZING INTERNATIONAL INFORMATION SYSTEMS

Optimizing international information systems depends on a sound intercultural theory of mind. The best method for this is to begin with the computational theory of mind, which is based on the concepts of computation and specialization. This means that conceptual modules are dedicated to specific tasks, such as visual processing. Culture is another area of the mind that occupies a conceptual module, which forms the basis for deep understanding through core values and beliefs. It is this intercultural theory of mind that is crucial for optimizing international information systems, because it affects everything from labels and categories to search queries and results.

Information Systems

Information systems are developed with the assumption that they serve some purpose for a user. The general properties of an information system include the following: focus, inclusiveness, completeness, structure, technology, and utility (Buckland, 1991).

Focus requires that a system support a particular definition and trajectory of the deliverable. Inclusiveness broadens the intention of focus, providing sufficient generalization to contextualize the focus. Completeness comprises a system's unifying characteristics, such as the synchronization between focus and inclusiveness. Structure refers to the architectural sensibility of the system, or the relations between parts. Technology refers to the online constraints of a system, such as platform and bandwidth limitations. Finally, the notion of utility justifies the investigation of optimization techniques. While the scope and goal of an information system address its practical implications, rigid optimization for international audiences is limited to large texts.

Information systems are designed according to the developer's classification strategies. Classification is rooted in the way items are labeled, and labels influence an online system's overall structure. And this means that an information system's taxonomy is critical for defining the user experience.

Taxonomic Structures

Classification is crucial for understanding complex international online factors, as it provides a useful extra dimension for conceptualizing the complexities of information systems. The classification of things reflects cultural ways of thinking. Classification also provides a conceptual and ontological bridge between a computational theory of mind, culture, and optimization.

A major distinction in the study of categorization is offered through monosemous and polysemous differences, a concept made explicit in the field of linguistics. Monosemous approaches toward categorization are rather similar to

Aristotle's classical system, which is based on deep analytics (Aristotle, 2001a, 2001b, 2001c). Lexical items either fit neatly within a given category or they do not, as realized through the principle of deduction (cf. Quine, 2006). Through determining what a subject is *not*, an online taxonomist becomes capable of appropriately filing a lexical item in an appropriate category—discrete classification. Examined through a vertical lens, monosemous classification constrains lexical items within pure hierarchical online nodes (Morville, 2005; Rosenfeld & Morville, 2002). Monosemous classification is also frequently associated with simple vocabularies, which aid in the rigorous forms of deductive logic (see Table 2).

The principles driving polysemous classification are different from those of monosemous prototypes. For example, polysemy is generally associated with dialecticism. Although dialecticism is often associated with classical Greece and the Socratic method, in the study of culture and classification it refers to a holistic perspective of analysis. Thus, polysemy is exemplified by its dialectical, multilayered, rich, and inclusive approach toward capturing lexical items. The perception of continuity, or continuous classification, reflects a blended and complex organizational sense (cf. Edelman, 2005).

The notion of imperfect induction further exemplifies polysemy. Once they are examined through a vertical lens, polysemous strategies organize lexical items across layered gradients or polyhierarchies (Morville, 2005; Rosenfeld & Morville, 2002). Thus, nodes or clusters of content or lexical items assume stochastic rather than discrete patterns (hence, polyhierarchies). Finally, polysemous classification is generally associated with complex vocabularies, as in the use of metaphor.

Optimization requires negotiating the complexities of all topics in an online system. Consider the relationship of various bird gradients to the category BIRD. Assume that warbler, owl, penguin, and hummingbird encompass the whole category, and that optimization of the information system depends on correctly projecting the audience's assumptions about the prototypical status of each lexical item. Although warbler is generally the prototypical member of the category BIRD (which is borne out in experiments), certain cultures report

Table 2. A Heuristic Describing Major Classification Differences Between Monosemous and Polysemous Taxonomic Structures

Monosemous	Polysemous
• Analytical	• Dialectical
• Discrete classification	• Continuous classification
• Pure hierarchy	• Polyhierarchy
• Simple vocabulary	• Complex vocabularies

different prototypical members; this is exactly what is found in the Sonoran, Mexico biosphere (Pinker, 2003).

In Sonora, the term hummingbird offers a higher profile depiction for all members of the category BIRD. If an online system offers an introductory portal, then categorical prototypes may be critical for optimization. Likewise, one may presume that inhabitants of the southern cone in South America may warm to the idea of a penguin as the prototypical member of the category BIRD, despite the warbler's exemplary position in various encyclopedias and dictionaries. Both examples—Sonora and the southern cone—indicate a preference for pure hierarchies where distinct lexical items comprise their own conditional clusters (cf. Morville, 2005).

Optimization through polyhierarchies occurs through gradient and transcendent relationships between lexical items. The warbler and owl, for instance, may occupy the same category if the target audience entails users in the arctic. Although warblers occupy a central position within the category BIRD, some users may complicate its status as a prototypical member.

Another example of optimization through polyhierarchical taxonomies is also found in Sonora, Mexico.[1] Based on the data collected in a study, it became clear that optimization through polyhierarchical structures was capable of accommodating the rich and varied expressions relating to nature through the use of metaphor. Sonoran participants, fluent in both English and Spanish, indicated a complex taxonomy of flora and fauna, all of which benefited from nonconscious, polysemous, and polyhierarchical taxonomies.

Consider the Spanish word for hummingbird, *colibrí,* which is rarely used in Sonora. The rich and abundant use of metaphorical terms permeates the everyday use of language in Sonora. A hummingbird is not a *colibrí.* A hummingbird is a *chuparrosa, chupaflor, picaflor, chupamirto de fuego,* and *joyas voladores.* All of the Sonoran terms for hummingbird express the vibrant and descriptive features of the bird's behavior. Especially interesting is the first term in the list, *chuparrosa,* which is a homonym, meaning both a hummingbird and a nectar-producing flower. Thus, *chuparrosa* is the primary term for a specific nectar-producing flower along the Sonoran corridor, serving a dual linguistic purpose in Sonora.

Clearly, the taxonomic doublenesses of Sonoran plants and animals—as in the symbiotic case of a hummingbird and a nectar-producing flower—present an internationalization team with structural problems. But if an information system is to be optimized for Sonoran users, a polysemous taxonomic structure is in order.

[1] Data were collected from two international environmental organizations, one in the United States and one in Mexico. The goal of the study was to determine whether any adaptations were required in optimizing an online module about local plants and animals. Reclassification became an important cultural adaptation in the project.

For Sonoran residents, a *chuparrosa* is both a bird and a flower, so the modularity should reveal the symbiotic relationship between the two in the natural world. This means that proximal navigation cues should reflect and reinforce related topics and terms (Rosenfeld & Morville, 2002). The hummingbird cluster should contain accessible links to the node *chuparrosa* for the nectar-producing flower.

Such an optimization is really quite obvious, but note that this example is based on a simplified use of polysemous and polyhierarchical patterns. A complex information system (such as the 9,000+ avifauna lexical items and their data) presents far more immediate complications than the example above, which is far simpler than many information systems. Truly complex international information systems are ruled not by the statistical curve of normal distribution but by power law logarithms in which the majority of links are connected by a small number of nodes (Buckland, 1991).

An international information system is optimized when it can capitalize on a power law to shape an efficient taxonomic structure, also known as a phase state. But taxonomic structure is only one feature of an information system. A user needs to find useful information, and this means not so much locating the correct content as eliminating unnecessary results (Buckland, 1991; Morville, 2005). Without an adequate retrieval algorithm, an information system is almost useless.

Retrieval Algorithms

An algorithm is a finite procedure designed to perform some function within a given interval of time (Franzén, 2005). Algorithms are crucial for information systems, as they are the engines that drive the process of retrieving, sorting, and displaying information. A search query, for example, operates through a finite set of procedures. Algorithms usually process data and information through iterative steps using decision rules based on logical operators.

Although retrieval algorithms are developed in a computer language, their effectiveness relies on natural language compatibility with the target audience. An algorithm is ineffective if the code driving the function does not use language that is semantically appropriate for the target audience. Thus, information systems for international audiences require an overhaul of the retrieval mechanism, a feature optimized through appropriate language use.

Sonorismos, or Sonoran sayings, offer a useful perspective on language and its implications for information retrieval. Consider the relationship between culture and language. Discourse features of Mexican Spanish include "flowery, poetic language" and a "flexible sentence structure," both of which contribute to the complexity of texts (Montaño-Harmon, 1991). The syntax of Sonoran Spanish tends toward the "additive, explicative, and resultative," to which are added numerous digressions (Montaño-Harmon, 1991). Sonoran Spanish is qualitatively different from American English, which relies on clarity and concision (cf. Williams, 2005).

Suppose one is developing a section of an information system dedicated to New World birds, such as the hummingbird.

Lexical nouns for Sonoran birds are often conceptualized through flowery and ornate constructions, with a strong tendency toward metaphor. The Spanish word for hummingbird is based on the Latinate *colibrí,* which means, literally, hummingbird, but few people in Sonora use the literal term. Rather, Sonoran residents frequently rely on hummingbird metaphors. A *colibrí* is a *chuparrosa* (sucker of roses and myrtle), *chupaflor* (flower sucker), *picaflor* (flower piercer), and *chupamirto de fuego* (myrtle sipper, color of fire). The *colibrí* is even a *joyas voladores*, or flying jewel.

Each of these Sonoran sayings indicates a stylized approach toward the use of language for the natural world. Particularly interesting among the Sonoran metaphors is the lexical item *chuparrosa*, or sucker of roses and myrtle. While *chuparrosa* is a general term for hummingbird, its name is originally associated with a nectar-producing plant—the *Justicia californica*. And this is a startling example of lexical cross-identification, which points to two entirely different though symbiotic organisms, their relationship based entirely on food and pollination.

But *chuparrosa* is not the only example of metaphor in Sonora. The dung beetle is also conceptualized as a football player and a shoeshine boy, the roadrunner as a countryman, and the dragonfly as a little cigar. Explaining the use of metaphoric Sonoran sayings as due to a close relationship with the environment is inadequate, as native speakers of English (who are also fluent in Spanish) are either unaware of the complex lexical terms along the Sonoran corridor or dismiss their relevance.

Retrieval algorithms often depend on some kind of semantic relationship operating in the background, as in a relational database. Semantic relationships differ both horizontally (being simple and equivalent) and vertically (being complex and associative), and all of these are described in terms of controlled vocabularies (Rosenfeld & Morville, 2002). A controlled vocabulary is any defined subset of natural language that consists of a list of preferred terms (Rosenfeld & Morville, 2002).

Consider the query term BIRD. A query based on a controlled vocabulary of simple equivalence offers no particular prototype for the category BIRD, because all lexical items are rendered the same. An algorithm for the term BIRD may locate, sort, and display warbler, owl, and penguin before hummingbird, which is problematic in Sonora. Clearly, the term penguin has no relevance in Sonora, yet a simple and equivalent controlled vocabulary may yield an irrelevant result. And an irrelevant query result results in a much larger chance of losing a user than just about any other kind of online factor (Morville, 2005).

Reconsider the query term BIRD. A query based on a vertical controlled vocabulary, through complex lexical association, is capable of retrieving terms based on semantic relevance for the target audience. A retrieval algorithm for the term BIRD, if adapted for Sonora, will place the term hummingbird first, followed by owl, warbler, and penguin. The term hummingbird connects prototypical lexical items and related information with users in Sonora, while the term penguin offers no conceptual relevance. Again, the simple and equivalent versus complex and associative retrieval algorithm optimization is simplified. And while this example is rudimentary, the impact of even a fairly minor change such as the change in placement of the terms hummingbird and penguin offers considerable insight into the complexities of optimizing an information system for international audiences.

Search engine optimization is the most important tool developers can use to connect information with their target audience (Morville, 2005). The average information system user, in English-speaking locales, spends fewer than 10 seconds browsing the results of a retrieval algorithm (Morville, 2005). The likelihood of rejecting the system is substantially raised if the lexical items of the query term are not adapted for the target user. Further, the majority of users never browse retrieval results beyond the second page (Morville, 2005). Even simple conceptual differences between related lexical items—such as the hummingbird, warbler, owl, and penguin—complicate a search query return.

CONCLUSION

Thatcher (1999) and St.Amant (2005) have been the two strongest proponents in the field for making changes in online internationalization. Just a few of the topics addressed by these authors include cultural and rhetorical adaptations, image redesign, issues of translation, outsourcing and offshoring, and accommodating the unique logistical problems of international collaboration. Clearly, the work of these scholars has done much to enrich our understanding of why designers and developers should reconsider the conventional work of internationalization. Further, their work has laid the foundation for scrutinizing the deeper aspects of the ways in which users from around the world interact with online computing.

In *Central Works in Technical Communication* (Johnson-Eilola & Selber, 2004), various authors assess the field through many different perspectives, which include historical, rhetorical, philosophical, ethical, methodological, workplace, online, and pedagogical. The book is an excellent contribution, containing both rich theoretical and clear practical advice. And while the book is considered a foundational text, its timeliness shows. For instance, Bernhardt's (2003) chapter about online text is now dated, which is and inevitable consequence of writing about online technologies. Despite its brilliance of the time, Bernhardt's article provided no insights into how international users should be reconsidered. This is somewhat surprising considering the author's longstanding international work at the time of its first printing, the ideas of his article were certainly relevant. The online environment is now dynamic, and this new trait requires much more of the people who plan to work in the field.

Culture takes on a more theoretical perspective in other parts of the book. Much is said about power and politics, and these two characteristics are clearly very important. But sometimes the best solution is from inside, and that means working within the constraints of a problem. Some areas of the globe surely have difficulty accessing online information, although this in itself cannot be remedied overnight. But what can be righted is the inclusion of crucial cultural values, the kinds of automatic beliefs that people use to make fast and frugal decisions. Other authors have also brought important insights into the internationalization process.

Other authors have also provided important insights into the internationalization process. According to Giammona (2004), the future of technical and scientific communication depends on an ability to see the big picture, an aptitude for instructional design, writer responsibility (or user orientation), and an ability to quickly grasp complex information. Giammona indicates the importance of international factors. The problem, however, is that the deepest dimensions of culture are never involved in the internationalization process. For that, we need information developers who have an intuitive grasp of the target audience.

Optimizing international information systems requires not only a reappraisal of culture but also a reworking of the structural and retrieval mechanisms that drive an online system. The process of optimizing an international information system is complex, but it must begin with a firm understanding of the computational theory of mind. The adaptive unconscious, which is the mind's autopilot, houses the deep values and beliefs that distinguish cultures. Culture is an important dimension to consider, because it accounts for the different labeling, classifying, and retrieval strategies found throughout the world.

But scientists still do not know how the mind works (Edelman, 2005; Pinker, 2005), intercultural scholars often lack the training in psychology or sociology necessary to understand the full relevance of their work, and information systems analysts continue to wrestle with multiple technical constraints. Nevertheless, it is almost impossible to develop an effective international information system without a working model of the ways in which different minds interact with the world. And this is crucial for optimizing international information systems.

REFERENCES

Abramov, I. (1997). Physiological mechanisms of color vision. In C. L. Hardin & L. Maffi (Eds.), *Color categories in thought and language* (pp. 89–117). New York: Cambridge University Press.

Aristotle. (2001a). Categories. In R. McKeon (Ed.), *Basic works of Aristotle* (pp. 7–39). New York: Random House.

Aristotle. (2001b.) Metaphysics (Zeta). In R. McKeon (Ed.), *Basic works of Aristotle* (pp. 783–811). New York: Random House.

Aristotle. (2001c). Rhetoric. In R. McKeon (Ed.), *Basic works of Aristotle* (pp. 1325–1454). New York: Random House.

Aykin, N. (2005). Overview: Where to start and what to consider. In N. Ayken (Ed.), *Usability and internationalization of information technology* (pp. 3–20). Mahwah, NJ: Erlbaum.

Aykin, N., & Milewski, A. E. (2005). Practical issues and guidelines for international information display. In N. Ayken (Ed.), *Usability and internationalization of information technology* (pp. 21–50). Mahwah, NJ: Erlbaum.

Bernhardt, S. (2003). The shape of text to come. In J. Johnson-Eilola & S. A. Selber (Eds.), *Central works in technical communication*. New York: Oxford University Press.

Buckland, M. (1991). *Information and information systems*. New York: Praeger.

Damasio, A. (2005). *Descartes' error: Emotion, reason, and the human brain*. New York: Penguin Books.

Davidoff, J. (1997). The neuropsychology of color. In C. L. Hardin & L. Maffi (Eds.), *Color categories in thought and language* (pp. 118–134). New York: Cambridge University Press.

Davidson, D. (2001). *Subjective, intersubjective, objective,* Cambridge, MA: Belknap Press of Harvard University Press.

Edelman, G. (2005). *Wider than the sky: The phenomenal gift of consciousness*. New Haven, CT: Yale University Press.

Forbes, N. (2005). *Imitation of life: How biology is inspiring computing*. Cambridge, MA: MIT Press.

Franzén, T. (2005). *Gödel's theorem: An incomplete guide to its use and abuse*. Wellesley, MA: A. K. Peters.

Giammona, B. (2003). The future of technical communication: How innovation, technology, information management, and other forces are shaping the future of the profession. *Technical Communication, 51,* 349–366.

Goldberg, E. (2002). *The executive brain: Frontal lobes and the civilized mind*. New York: Oxford University Press.

Goldman, L. et al. (1996). Prediction of the need for intensive care in patients who come to emergency departments with acute chest pain. *New England Journal of Medicine, 334*(23), 1498–1504.

Hofstede, G. (2005). *Cultures and organizations: Software of the mind*. New York: McGraw-Hill.

Johnson-Eilola, J., & Selber, S. A. (2004). *Central works in technical communication*. New York: Oxford University Press.

Marcus, A. (2005). User interface design and culture. In N. Ayken (Ed.), *Usability and internationalization of information technology* (pp. 51–78). Mahwah, NJ: Erlbaum.

Montaño-Harmon, M. (1991). Discourse features of written Mexican Spanish: Current research in contrastive rhetoric and its implications. *Hispania, 74,* 417–425.

Morville, P. (2005). *Ambient findability*. Sebastopol, CA: O'Reilly.

Pinker, S. (2003). Language as an adaptation to the cognitive niche. In M. Christiansen & S. Kirby (Eds.), *Language evolution: State of the art* (pp. 16–37). New York: Oxford University Press.

Pinker, S. (2005a). The faculty of language: What's so special about it? *Cognition, 95,* 201–236.

Pinker, S. (2005b). So how does the mind work? *Mind and Language, 20*(1), 1–14.

Quine, W. V. (1998). *From stimulus to science* (Rev. ed.). Cambridge, MA: Harvard University Press.

Quine, W. V. (2006). *The philosophy of logic* (2nd ed.). Cambridge, MA: Harvard University Press.

Rosenfeld, L., & Morville, P. (2002). *Information architecture for the World Wide Web.* Cambridge, MA: O'Reilly.

Seife, C. (2006). *Decoding the universe: How the new science of information is explaining everything in the cosmos, from our brains to black holes*. New York: Viking Penguin.

St.Amant, K. (2005). A prototype theory approach to international image design. *IEEE Transactions on Professional Communication, 48*(2), 219–222.

Taylor, J. R. (2003). *Linguistic categorization: Prototypes in linguistic theory* (3rd ed.). New York: Oxford University Press.

Thatcher, B. (1999). Cultural and rhetorical adaptations for South American audiences. *Technical Communication,* second quarter, 177–195.

Todd, P. M., & Gigerenzer, G. (2000). Précis of simple heuristics that make us smart. *Behavioral and Brain Sciences, 23,* 727–780.

Williams, J. M. (2005). *Style: Ten lessons in clarity and grace* (8th ed.). New York: Pearson Longman.

Wilson, T. D. (2002). *Strangers to ourselves: Discovering the adaptive unconscious.* Cambridge, MA: Belknap Press of Harvard University Press.

CHAPTER 4

Cyberspace, International Intellectual Property Law, and Rhetoric

Martine Courant Rife

CHAPTER OVERVIEW

In this chapter, I argue that for technical communicators working in cyberspace, rhetoric has new relevance because of its ability to assist strategic thinking; and that a basic awareness of international intellectual property (IP) law is crucial to the success of the field of technical communication, since laws often (invisibly) shape communication and culture as these two concepts interact in cyberspace and in law. Specific examples of international IP law's influence on technical communication are taken from legal cases, international definitions of *originality,* and recent issues surrounding geographical indicators. I conclude by arguing that increased knowledge of these issues will benefit the field of technical communication by enabling technical communicators to add value to their organizations via strategic thinking.

I have been studying intellectual property law and its intersection with digital-technical communication for over 3 years. My research has necessarily taken me into the realm of international law, because cyberspace challenges our traditional understandings of geographically based place and ties to geographically-based law. Writing in cyberspace presents exceedingly complex issues of law, as the texts we create truly circulate in international spaces and in spaces that we cannot even imagine. This chapter attempts to give a broad overview of some key issues that arise at the intersection of international law and writing in cyberspace, under a rhetorical lens. In this chapter I make two arguments. One, that for technical communicators working in cyberspace, rhetoric has new relevance because of its ability to assist strategic thinking; and two, that a basic awareness of international intellectual property law is crucial to the success of the field, since laws often (invisibly) shape communication and culture as those two concepts interact in cyberspace and in law. I discuss international law[1] in a rhetorical framework informed by Kenneth Burke's (1969) concepts of *identification* and *division.*

In John Logie's book, *Peers, Pirates, and Persuasion* (2006), he folds Kenneth Burke's identification into a discussion of Napster's carefully constructed cyberspace-ethos. Napster was the maker and distributor of peer-to-peer file-sharing software; this software was ultimately found to be in violation of U.S. copyright law. As Logie points out, "Burkean identification offers rhetor and audience the opportunity to unite over a shared principle without necessarily verifying the virtue of either party" (2006, p. 41). Logie's use of Burke's identification in the context of intellectual property and cyberspace is helpful and worthy of further exploration. In *A Rhetoric of Motives*, Burke (1969) states that in order to have identification, there must be division. One cannot identify with another unless one is already divided from that other:

> A is not identical with his colleague, B. But insofar as their interests are joined, A is *identified* with B. Or he may *identify himself* with B even when their interests are not joined, if he assumes that they are, or is persuaded to believe so. (p. 20)

According to Burke, identification compensates for division, but in turn, it is my position that division also compensates for identification, as illustrated by some cyberspace-law examples provided below.

Burke places a discussion of property in the context of his discussion of identification. A thing is identified by its properties, and as Burke asserts, in the realm of rhetoric, "such identification is frequently by property in the most materialistic sense of the term, economic property" (pp. 23–24). This raises issues surrounding intellectual property, the "creations" of one's mind,[2] as it circulates in cyberspace. The rhetorical terms identification and division as Logie hints, serve as a useful framework for discussing cyberspace and law. Cyberspace, the place and the act, provides users and participants with a space of identification and a space for communal action. International intellectual property laws provide for identification in cyberspace, but reactive localization of those laws, in practice, produces division.

[1] Notes begin on p. 99.

Starke-Meyerring (2006) argues that because of globalization, professional communicators should learn to understand blurred boundaries of citizenship, geographical ties, and cultures; we have a "new urgency . . . to shift our understanding of culture away from specifiable, fixed boundaries; lists of traits ascribed to entire groups of people" (p. 479). Yes, the World Wide Web (WWW) facilitates identification and in some cases presents a platform where people can negotiate and renegotiate identities and cultural identification, but it appears this negotiation can only go so far from geographical locations before it returns again to these kinds of material, land-based connections. Division arises in the context of cyberspace, or is reinscribed, at least in part because division compensates for identification. This is a trend. For example, after the nation-state Sierra Leone crumbled in the 1990s, members of its community kept the nation alive through conversation on a list serve called Leonenet. Cyberspace offered them an opportunity to identify as a virtual community, even a virtual nation, but they distinguished/divided themselves from other virtual communities by retaining their local, place-based identities. "Even though the internet appears to be a deterritorialized communicative space, Halavais (2000) has shown that national borders do appear on the world wide web" (Tynes, 2007, p. 499). In its mission statement, Leonenet says it is for "Sierra Leoneans and friends of Sierra Leone" (Tynes, 2007, p. 503). According to Tynes's definition, a virtual nation cannot be such unless it is involved in the process of building/rebuilding a nation. "Leonenet can be viewed as a space where Sierra Leone's state-related symbols are generated and then held in conceptual escrow, waiting for the moment when the state returns to Sierra Leone" (p. 502). Tynes's discussion of the virtual-nation Sierra Leone is grounded in a notion of cyberspace as a virtual-space inviting rhetorical invention as outcome to the imagined, or imagination. While Sierra Leone did not exist as a geographical location, it was discussed by the list serve members as if it really did exist, just as architects working on a building discuss and imagine the building as if it were an "already present reality," causing the building characteristics to become increasingly concrete (Haller, 2000, p. 379).

Burke illustrates with an extreme example how division arises from cooperation via identification: "in the end, men are brought to that most tragically ironic of all divisions, or conflicts, wherein millions of cooperative acts go into the preparation for one single destructive act. We refer to that ultimate *disease* of cooperation: *war*" (1969, p. 22). An understanding of this might assist the technical communicator with strategic thinking, especially when innovation in cyberspace is involved. A technical communicator who is aware of the interaction of identification and division in legal-cyberspace contexts (the state, for example, is at least in part constituted by law), and who has familiarity with rhetoric so as to be able to carefully weigh various probabilities, will add value to his/her organization. In cyberspace, identification and division are placed ambiguously together in such a way that one cannot easily tell where one ends and the other begins; in such circumstances "you have the characteristic invitation to rhetoric" (Burke, 1969, p. 25). In my examples below, I attempt to separate out ways in which users rhetorically construct their own division from larger communities—by naming and defining certain terms as acts of implementation of intellectual property law.

Technical communication has longtime ties to rhetoric and more recent ties to intellectual property issues. Brockmann (1988, 1998, 1999) studied the ways in which inventors used technical communication and rhetoric to achieve recognition through the U.S. patent system. Bazerman (1999) continued this work by examining Thomas Edison's speech acts as inscribed in his notebooks, paying particular attention to how patent law influenced the writing of Edison's technical documents. Durack (1997, 2001) argues that technical communication researchers might study the patent record in order to trace the history of the discovery and development of technical communication. She discusses the relationship between patent systems and the peer review process in order to illustrate how university patent activity influences the production of scientific knowledge. Finally, Durack (2006) points out that as current trends continue, the researcher must acknowledge the presence and role of the patent system on the creation of scientific knowledge; her research points directly to the increased need for technical communication scholars to attend to the ways in which the law and legal infrastructure shape their own writing practices and the creation of knowledge.

A recent publication shows that globally, the intersection of law and technical communication is a general concern for both practitioners and academics (Hennig & Tjarks-Sobhani, 2004). Issues of how law influences the creation of knowledge and texts in digital-legal contexts are addressed in scholarship by DeVoss and Porter (2006), Gurak (2005), Herrington (1997, 1998, 2001, 2003), Howard (2004), Porter and Rife (2005), and Rife (2006a, 2006b, 2007), among others.[3] With the advent of remix culture, "new media" compositions, and globally, digitally connected writing environments, that is, cyberspace, issues of copyright/fair use in technical writing practice have become central. Very little attention has been given to international law, cyberspace, and technical communication. Yet, technical communicators voice concern for law-related issues such as the increasing number and complexity of laws, the repercussions of not understanding laws, the quality and usability of technical documentation versus its legality, and the lack of uniform application of (Hennig & Tjarks-Sobhani, 2004).

In the Hennig and Tjarks-Sohani (2004) collection, technical communicators from across the globe provide their perspectives. With regard to concern over increasingly complex laws and their impact on technical documentation, Giliarevski (Russia) lists seven Russian laws that regulate documentation, including electronic documentation (2004, pp. 136–137). Voichita-Alexandra Ghenghea (Romania) discusses how Romania's government ordinances set out specific requirements for documentation. Technical documentation in Romania must inform the consumer fully, correctly, and precisely, must include foreseeable risks, and must be written in Romanian. Kristina Abdallah et al. describe Finland's legal framework under the European Union (EU). Members of the EU, such as Finland, are bound by EU Directives. which attempt to effect harmonization among members, but Finland also has its own set of laws. Abdallah et al. (2004) list five Finland-specific laws that are layered over EU Directives. Additionally, Finland requires instructions associated with consumer goods to be written in both Finnish and Swedish. Other concerns expressed by authors of essays in the collection include concerns about the legality of documentation versus its usability, as well as the repercussions of not understanding legal requirements. Verhein-Jarren (Switzerland) notes that while legal requirements strengthen technical documentation, they also weaken it, because as writers tailor documentation to meet specific laws, "the requirements of a user are nowhere in the picture" (p. 25). Like Verhein-Jarren, Giliaresvski (Switzerland) refers to the laws' failure to address "all interests of writers and users" (2004, p. 137). Peter Ring (Denmark) notes, "I don't think quality control will be common before somebody is heavily fined" (2004, p. 64). Authors representing an array of countries understand the complexity of laws that shape technical communication and are concerned about user/consumer needs in these contexts.[4]

International technical communication should be concerned with lack of uniformity in the application of laws. Legal issues that impact technical communication have evolved and been localized by different countries. Global uniformity in laws is probably not possible because global uniformity, or identification, would subsequently invite division, brought about by the localization of laws for different jurisdictions. This is problematic in cyberspace environments, because one might create a text in jurisdiction A pursuant to that jurisdiction's laws, and have that text circulate in jurisdiction B, a place that has conflicting laws. Will a technical communicator or an international organization in this circumstance have liability for a text that is "legal" in one jurisdiction but turns out to be "illegal" in another jurisdiction? It depends. Such a determination would involve the presence or absence of international agreements plus the strength of the ties between the technical communicator and jurisdiction B. Sufficient ties, or the presence of international agreements, could cause the technical communicator or organization to be liable. In *Gilliam v. American Broadcasting Companies, Inc.* (1976), "Monty Python" (that is, writers and performers from the television series *Monty Python*) sued American Broadcasting (AB) for misrepresenting Monty Python's work. AB aired an edited version of the programs. These programs were originally broadcast in their entirety by the BBC, but because of time limitations due to the insertion of commercials, AB severely edited Monty Python's work, with the result that Monty Python (MP) was "'appalled' at the discontinuity and 'mutilation' that had resulted" (*Gilliam v. American Broadcasting,* 1976, p. 18). MP argued a "moral rights"-based claim,[5] that AB impaired the integrity of MP's work in such a way that the damage was irreparable. The problem was that in the United States, there is no moral rights regime fashioned to protect integrity of work outside the parties' contract (other than the limited protections offered by the Visual Artists Rights Act). Nonetheless, although the main arguments in the case were based on copyright infringement, the court fashioned legal redress for MP based on U.S. trademark law's prohibition or misrepresentation. In a roundabout way, the U.S. 2nd Circuit Court of Appeals allowed British moral rights protection for Monty Python by using a clever reinterpretation of U.S. trademark law.

Another example of one country's law crossing over into another country, is seen in a French court case, *Turner Entertainment Co. v. Huston* (1994). The case concerned John Huston's film *Asphalt Jungle*, produced by MGM in the United States in 1950. The movie was filmed in black and white by American John Huston, and a U.S. copyright was obtained that same year by Huston's employer, Lowe's Inc. The registration was renewed in 1977 and transferred to Turner Entertainment Co. in 1986. In 1988, Turner had the movie colorized and scheduled a broadcast on June 26, 1988, on the French Fifth Television Channel. Before the movie was aired, a lawsuit was filed in France's court system by John Huston's heirs (Angelica, Daniel, and Walter), among a number of other interested parties. "They opposed the broadcast because they deemed it a violation of the author's moral right, aggravated in their opinion by the fact that John HUSTON had opposed colorization of his works during his life" (*Turner v. Huston,* 1994, p. 1). Citing the Universal Copyright and Berne conventions, the French court stated that "the film's colorization without authorization and control by the authors or their heirs amounted to violation of the creative activity of its makers, even if it should satisfy the expectations of a certain public for commercially obvious reasons; the use of this process . . . infringed the moral right of the authors as mandatorily protected under French law" (p. 6), ordering Turner to pay fines to Huston's heirs, and preventing the broadcast of the colorized version. One wonders how these kinds of issues might play out in cyberspace. The United States does not recognize moral rights, but content created and modified in the United States and circulated in countries that do have these kinds of protections could theoretically trigger such liability. The possibilities at the present moment are open, and should be considered by technical communicators working in international environments. Awareness is necessary.

While it is commonly believed that there are international laws that apply uniformly to each country, this is a fallacy. Instead, different groups of sovereign entities or nation-states agree to try to agree on some international standards: identification. In order to preserve division via sovereignty and autonomy, each individual country creates a unique method of implementation of an international standard. The implementation is subsequently subject to a country's interpretation. Interpretation is in turn subject to a country's history, culture, values, and to a large extent, power in the global economy. More powerful countries have more leverage in localizing and customizing implementation of international standards/agreements than do less powerful countries.

Rosemary Coombe (2005) argues that cyberspace and "relations of transnationalism are enabling new diasporas to simultaneously spread and reinforce forms of national cultural difference that are no longer (if they ever were) congruent with state territorial borders even as these 'nations' are given new meanings as they are experienced and interpreted in the places of others" (p. 5). Tynes's (2007) case study of Leonenet confirms this. Coombe (2002) states that alliances have been formed by nongovernmental organizations (NGOs) around the world, "including First Nations peoples, Maori, Australian Aboriginal and Torres Strait Islanders, Andean indigenous people, Sami peoples, and Ainu . . . [people of the] Mongolian steppes, the Solomon Islands, India, and Bangladeshi hill tribes . . . descendants of escaped slaves in the Americas" (p. 277). Internationalization and cyberspace make communication easier, even possible, for indigenous peoples, and other communities of persons not associated with one "place," such as the members of the diaspora from Sierra Leone who gathered together via the Web. These kinds of cyberspace gatherings for marginalized communities have had some positive legal-impact via global awareness of their plight. While cyberspace provides a place for identification, it also provides a place for division, or the bringing together of groups of "difference." Block's (2004) study supported the idea that while some anticipated that English would become the predominant language on the Internet, other languages such as Spanish, German, and Japanese have made robust spaces for themselves.

ORIGINALITY AS IDENTIFICATION AND DIVISION

Texts circulating globally in cyberspace have made intellectual property ownership issues more salient than ever. In the following extended example, I show how originality, a term important to defining authorship and key to dividing intellectual property rights, is both global and local simultaneously, thus supporting Burke's notion of identification and division. However, by rhetorical analysis, one can separate out the localization of this concept from the globality of it. The concept of originality is one that is important for technical communicators to understand in digital-international environments, because originality is an attribute required of texts that are permitted copyright protection in many, if not most, countries that are signatories to international agreements like the Agreement on Trade-Related Aspects of Intellectual Property Rights (TRIPS) and the Berne Convention. Copyright protection provides "property" rights to the holder of the copyright, thus reinscribing division. A copyright holder is different from a person who does not have those rights in a text.[6] Originality is key to the authorship function. Foucault (1984), predicting the author's death, characterized the old question as follows: who really spoke and with what *originality*? The post-author question is . . . how can it circulate and who can appropriate it for himself—what are the places in it where there is room for possible subjects, and who can assume these various subject functions? While international courts examine and define originality as an author-function question, in fact, after a closer look, the courts really answer the post-author questions. How can the text circulate—who can appropriate it? If a text is original, then it is copyrightable. If it is copyrightable, then others have less ability to appropriate it or occupy it than if the text is *unoriginal*. But what constitutes originality varies between countries. Canada has one definition, the U.K. another, and the United States a third. The technical communicator seeking copyright protection for a work, or wondering whether another's work is copyrighted, might consider these kinds of variations. Originality of work provides a type of identification—Canada, the United States, and the U.K. can identify with each other because each country requires originality in copyrightable work. But these countries are divided upon the exact meaning of originality—this is problematic in the context of creating and using texts in cyberspace.

A recent Canadian judicial opinion serves as a clear example of such identification and division in play. In *CCH Canadian Ltd. v. Law Society of Upper Canada* (2004), three publishers, CCH Canadian Ltd., Thomson Canada Ltd., and Canada Law Book Inc. (CCH), sued the Law Society of Upper Canada for copyright infringement (Rife, 2006b). The Law Society was established in 1822 and regulates the Canadian legal profession. It maintains the Great Library in Toronto, a research library containing one of the largest collections of legal materials in Canada. The Law Society provoked publishers by sending out copies of select texts and portions of those texts through a custom copying service to members, judges, and authorized researchers, and providing self-service copy machines in its Great Library for use by patrons. Ultimately, the Canadian Supreme Court held in favor of the Law Society, but in its deliberations, it had to consider whether copyright law was triggered at all, that is, whether the texts in question were protected; thus the court had to define originality.

Originality was an issue in *CCH* because the publishers argued that their works were protected by copyright, and Canada's Copyright Act permits copyrights for "every original literary, dramatic, musical and artistic work" (Section 5). Originality is defined in the act as including works in the literary domain as well as all other forms/modes of expression. The texts in question raised the issue of whether material such as headnotes, topical indexes, and compilations of reported judicial decisions were "original," such that copyright protection was triggered. To localize its definition of originality in an international context, the Canadian court explicitly looked to, that is, identified with, the definitions of originality of other countries: the United Kingdom, the United States, and France. Citing two cases from the United Kingdom and Canada (both used metrics of "industriousness," "sweat of the brow," and "time and labor"), the court also looked to the U.S. standard of originality per *Feist Publications Inc. v. Rural Telephone Service Co.* (1991). In *Feist,* the U.S. Supreme Court considered whether compilations manifested in a telephone directory were original—deciding that a compilation in this case was not, since it lacked the requisite "modicum of creativity": the new creativity standard fashioned by the court. After examining *Feist*, the *CCH* court reiterated that Canada traditionally had a very low bar for what constitutes original work (the industriousness standard or "time and labor"); the *CCH* court stated that in recent years the courts (both Canadian and the United States) had moved away from the industriousness standard, and that to hold to that standard in this case would privilege the rights of authors so much that the scales would be tipped too far in their favor and against the public need to maintain a robust public domain. The court defined originality to include skill and judgment rather than time and labor. It rejected the U.S. measure of creativity as per *Feist,* stating that both the industriousness standard and the creativity standard were "extremes." So in the opinion, the Canadian Supreme Court identified with U.S. law but also divided itself from that law—via rhetorical analysis, one can separate these issues out, useful skill for technical communicators working in cross-cultural cyberspace environments.

Through the language of the judicial opinion, a customized law was created. As illustrated in Figure 1, the *CCH* court positioned itself with respect to other paradigms of "originality," that of previous Canadian law, U.K. law (time and labor, sweat of the brow), French law (skill and judgment). and U.S. law (modicum of creativity). The *CCH* measure includes creativity, but it also defines more works as original if viewed on a continuum between the creativity measure and the sweat of the brow measure.

CCH states: "While creative works will by definition be 'original' and covered by copyright, creativity is not required to make a work 'original.'" Note that the *Feist* court does not mention Canada and does not look to any other country for guidance, which suggests that rhetorically, the U.S. judicial opinion may be attempting to provide a universal standard of originality for the rest of the world,

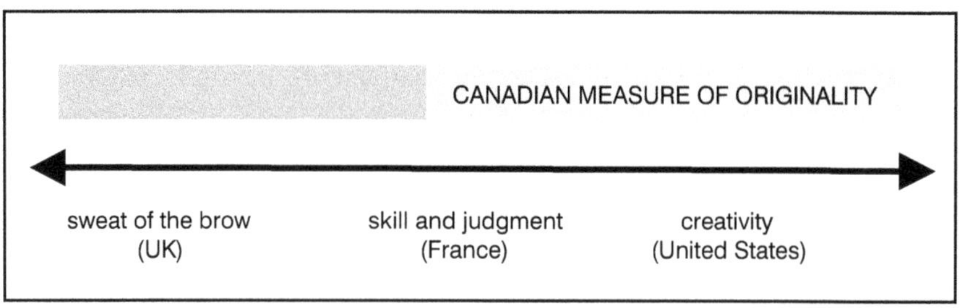

Figure 1. The Canadian originality metric on an international continuum. Originality serves as a point of identification, but also as a point of division.

against which less powerful countries might measure themselves. However, the Canadian court cleverly avoided doing this by looking to other countries' definitions of originality in addition to that of the United States. The *CCH* court tries to differentiate between *creativity* and its own standard of *skill and judgment,* finding that the texts in question, including headnotes, topical indexes, and compilations, were original and protected by copyright. While the court ended up defining originality as an author-function, it really did this by answering the question through the creation of a continuum that fits nicely into Foucault's proposed post-author questions: Where is there room for possible subjects, and who can assume these various subject functions? Based on the *CCH* continuum, possible subjects can be "authors" of original work under Canadian law, as long as they create based on measures including *at least* skill and judgment. Canada identifies with France and the United States, yet it is difficult to know how different the *CCH* metric for originality is from the *Feist* metric until we have another set of facts tested in the courts. The *CCH* opinion illustrates how an international concept like originality can be differentiated, contrasted, and analyzed for local purposes but continue serving as a term of identification. The *CCH* judicial opinion evidences rhetorical strategies employed—whether consciously or not—to assert and crystallize local identity in international contexts. Technical communicators who gain awareness of these types of identity/ cultural constructions in international-cyberspace contexts will be better able to think strategically when selecting and planning content for WWW spaces.

GEOGRAPHICAL INDICATORS: COMMODIFIED CULTURE AS RESISTANCE TO CYBERSPACE

Geographical indicators are a type of legally recognized and protected property, maintained through texts in order to rhetorically construct division while simultaneously inviting identification. Geographical indicators (GIs) can be considered a subset of trademarks, and are defined in Article 22(1) of the World Trade Organization's (WTO's) 1995 Agreement on Trade Related Aspects of Intellectual Property Rights (TRIPS) as "indications which identify a good as originating in the territory of a Member, or a region or locality in that territory, where a given quality, reputation or other characteristic of the good is essentially attributable to its geographic origin" (United States Patent and Trademark Office, 2007, p. 1).

Any international business can set up a Web page but how will businesses differentiate themselves from one another via the Web? One increasingly popular way to do this is to reinscribe the importance of ties to geographical locations—geography is of course all but "erased" in cyberspace. In the Leonenet example, list serve members identified with the geographical place where their nation-state had existed. Coombe (2002) points out that geography may be more important than ever as part of U.S. patent processes. She argues that to protect the traditional knowledge of mainly poor people in the world, U.S. patent applications should be required to disclose the geographical originals of the invention they wish to patent, thus giving advance notice to people whose livelihood is place based and vulnerable to exploitation and appropriation. Because of information disseminated via the Web, powerful entities can more easily appropriate economically valuable intellectual property (trademarks, patentable ideas, copyrighted material, geographical indicators) from less powerful entities. Understanding this enables technical communicators to think strategically and act accountably. Quoting Carolyn Miller and Dale Sullivan, Thomas Huckin (2002) argues for the importance of accountability for technical communication in light of globalized and international environments.

Geographical indicators are textual identifiers of the *physical* origin of a good or product: Florida oranges, Louisiana catfish, Bordeaux wine, Roquefort cheese, Vidalia onions. GIs are often legally protected under international agreements: this is legally protected place-based *division*. Indicators arguably protect quality and are important marketing tools in the global marketplace. Whether or not a consumer is persuaded that catfish grown in Louisiana are actually superior in quality to catfish grown in, say, Vietnam has a great deal to do with how Louisiana catfish are rhetorically constructed in cyberspace marketing contexts. Cyberspace supposedly removes geography from the mix, but in response, the importance of geography is reinserted. This power struggle is between sovereign countries. In the recent "catfish wars," this issue was illustrated. The Catfish Farmers of America lobbied the U.S. Congress for a declaration that of 2,000 catfish species, only the U.S.-born family of Ictaluridae could be called catfish. Congressman Mike Ross (2002) via post to his website used cyberspace to support this decision and to reiterate the difference between Ictaluridae and Vietnamese catfish.

Vietnamese producers were forced to market their fish in the United States under the Vietnamese names *basa* and *tra* (Lam, 2003). The Catfish Farmers of America became upset because Vietnam imports captured 20% of the $590 million market for foreign catfish, arguably by selling at below-cost prices. Vietnamese importers, on the other hand, argued that they could simply produce the catfish more cheaply because of lower costs. The simple act of naming in this case is said to have impacted the economy of an entire nation, and so when technical communicators consider international marketing issues and cross-cultural branding, an understanding of the repercussions of "simple" naming and international law may prove beneficial.

Globalization is the term coined to describe "strategies to build global advantage by exploiting local difference" (Jessop, 1999, p. 12). Strategies used to maintain division through constructing local identities, sovereignty, and autonomy, along with strategies to maximize place-based competition continue to surface in light of globalization and in light of the multiplication of international relationships.

Globalization in turn is tied to cyberspace and the exchange of information. Geographical indicators (GIs) have become more important recently because certain localities want to economically protect and exploit their differences under the threat of increasing cross-country information flows, the threat of universalism, or global homogeneity.

Cyberspace provides a space for identification and commonality. The division created by geographical indicators, as difference-proscribed-by-law, compensates. In light of this tug-of-war, technical communicators who understand that sovereign entities and/or nation-states and at least some of their residents (along with members of other communities based on race, class, sexual orientation, etc.) want to maintain their individuality, autonomy, and difference. They want division, as is apparent, for example, from the assertion of geographical indicators tied to branding of products. This is where rhetoric becomes important, because division and identification are constructed through means of persuasion—and one of those means is law. One of the underlying legitimacies of the Weberian state is law (Weber, 1947), and one of the three areas of state-society relationships that help constitute the state, according to political scientist Joel Migdal (2001), is law. Cyberspace will not erase the state—so the question is, since cyberspace potentially erases geography, how will the state reinscribe itself via law in cyberspace?

Geographical indicators provide a legal method for dividing, commodifying, and protecting culture and boundaries on the international level. The downside to legally creating such division is that culture is reduced to commodity. In order to address this problem at the international level, in 2001 the United Nations Educational, Scientific and Cultural Organization (UNESCO) issued a declaration attempting to give guidance on how to provide cultural diversity other than via legal commodification. Some member states of UNESCO found the 2001 declaration inadequate as a response to "specific threats to cultural diversity in the era of globalization" (Coombe, 2005); the Convention on the Protection and Promotion of the Diversity of Cultural Expressions was consequently adopted in October 2005 (Bureau of Public Information, 2005). Even so, the economic incentives for certain localities to develop distinctive cultures in order to attract tourists and business loom large. Rosemary Coombe (2005) describes a speaking engagement she had with artists at the Academy of Maastricht, who had been approached by officials from the City of Leuven (Belgium) seeking ways to create "urban distinction" (p. 10). The artists' project was the city's effort to develop a "cultural communication plan" (p. 10), part of which was leveraged in cyberspace.

The pull of cyberspace as a place for identification furthers the desire and need for division through identity branding even for more traditionally marginalized peoples who seek economic development in light of increasing liberal international trade and competition:

> Popular strategies involve reifying local traditions and burnishing images of rustic authenticity onto goods, some of which were marginal products only two generations ago. For example, in some Italian mountain areas, the creation of telleggio—a particular kind of cheese—has become central to local development strategies. (Coombe, 2005, p. 12)

International intellectual property law affirms division via branding by protecting GIs. This is problematic, because consumerism, accepted as the norm in many areas (such as the United States, the United Kingdom, and parts of Asia), goes against other cultural values: "according to West-African ideology, over-consumption and hoarding are often thought of as witchcraft (Comaroff & Comaroff, 1993) or socially destructive behavior" (Tynes, 2007, p. 513). Yet, for protection for geographically produced items to be secured, commodification of culture is required. Certain *methods* of producing a geographically branded product must be developed and then reinscribed in texts. For example, in the wine industry, in order to protect a geographical indicator associated with a type of wine, methods such as "the spacing of vines, the pruning methods, permitted forms of fermentation" are developed and self-imposed by producers in order to create brand distinction (Coombe, 2005, p. 13). The same is true with the cheese industry—these practices rely on the notion that the procedures cannot be replicated in other geographical locations, regardless of the fact that information can easily be exchanged in cyberspace. GIs serve as dividers in response to flattening, to globalization through cyberspace. The importance of GIs to an economy varies widely. The United States tends to find GIs less important, while developing countries often find them more important to protect. But because of cyberspace, the middleman is arguably unnecessary. Consumers can conduct direct interactions with producers. Therefore, the rhetorical construction of difference on the Web may be the only point of differentiation between products.

While GIs reinscribe division, they must also invite identification at some level. A consumer purchasing Louisiana catfish or fair trade coffee must see him/herself as desiring something "authentic" or as supporting "social responsibility." The construction of the marketing message must carefully weigh strategies of division against strategies of identification. Thatcher's (2001) research confirms that readers in differing geographical areas prefer writing that fits within their own rhetorical frameworks. In order to demonstrate the need for empirical research regarding communication practices in intercultural environments, Thatcher completed two English translations of a letter that was published in a daily newspaper in Guayaquil, Ecuador. Thatcher's "Letter O" was a close translation of the letter, and his "Letter R" was a "rhetorical" translation in which he networked in meaning that he thought relevant to U.S. readers based on his experience as a U.S.-born technical communication teacher. Thatcher describes the changes he made to Letter R as situating the letter more in written culture, and appealing to U.S. Americans' self-identification as individuals, pointing out that a number of researchers claim the United States to be the most individualist country of all (p. 471). Letter O, in contrast, better reflects what Thatcher describes as Ong's descriptors of oral culture, including material that U.S. readers might find "indirect, circumlocutory, and pretentious" (p. 465), but with which, Thatcher claims, the South American survey respondents identified.

Thatcher surveyed 200 U.S. Americans and 200 South Americans regarding their letter preference (Letter O or Letter R).[7] Of 200 U.S. Americans surveyed, 100% preferred Letter R, while 67% of the 200 South Americans surveyed preferred the close translation (Letter O), 21% were undecided, and 12% preferred Letter R (2001, p. 459). Thatcher concludes that comparing these two types of readers, U.S. Americans and South Americans, in this case generally yields unified results with sharp contrast, since all of the U.S. readers preferred Letter R, and most of the South American readers preferred the close translation, while those that did not still recognized certain rhetorical patterns in Letter O that corresponded to oral culture.[8] Like St.Amant (2002), Thatcher found that particular rhetorical strategies are not universally effective. The same is true of international laws; they are simply standards that are then taken and localized in favor of sovereign entities in a way that (ideally) suits their individual rhetorical and ideological framework. Since technical communication and its products (manuals, directions, instructions, Web interfaces) flow internationally, the localization of related laws makes it very complicated and difficult for any generalizations to be drawn about, say, the "correct" way to draft a technical document in light of international readers, or the "correct" way to draft a marketing message.

As Thatcher (2001) points out, in some countries, written laws have little agency. My own research supports this: China, for example, has copyright laws on its books that as strict as any others, and yet those laws have little agency (Fishman, 2005). In contrast, Emile Cloatre (2008) of the University of Nottingham, U.K., has published a study of patent law in Djibouti, finding a total disconnect between written law and practice. In this case, patented pharmaceuticals were widely and exclusively prescribed in Djibouti even though no written law existed mandating this practice. The reason is primarily, she found, that French pharmacists have a huge presence in Djibouti, and in France, pharmacists almost exclusively rely on patented pharmaceuticals rather than generics. What Cloatre found is rather stunning, in that we cannot assume a written law has anything at all to do with social practice. The law's agency will be impacted by cultural issues. Thatcher frames these issues as in part dependent on written versus oral culture or on individualist versus collective culture. Further research in this area examining the agency of laws is needed in our field particularly, because laws are often written, and in professional communication we care about the agency of written texts. Cloatre (2008) used actor network theory to uncover the motivations for cultural practices supporting the use of patented pharmaceuticals in Djibouti.

In his dissertation, St.Amant (2002) finds that international online communication can suppress marginalized voices, and that "deep seated cultural values might not be open to compromise" (p. 4). In his research, St.Amant studied the rhetorical turns taken in an online discussion board when the EU Data Protective Directives (privacy provisions) and the U.S. Safe Harbor Principles were being adopted. Instead of adopting a version of the EU provisions, the United States adopted the Safe Harbor Principles. On the discussion board that St.Amant researched, there were two basic levels of communication: communication between interests in the United States, attempting to get the United States to take a certain approach in international negotiations, and on another level, communication between government officials from the United States and the EU. Membership in international organizations is used to evaluate cultures, and because countries/sovereign entities, and cultures will want to maintain autonomy, membership in international organizations or being signatories to treaties, such as TRIPS or Berne, will produce international standards. Standards serving as a point of identification will then be localized in a particular geographical location (this location may be more metaphorical than fixed in the case of sovereign indigenous peoples), just as the U.S. Americans and the Chinese in Sun's (2004) study localized the genre of text messaging. This reinscription of difference or division through localization will continue to be emphasized in online environments. As various cultures are exposed to other ways-of-being and ways-of-knowing. the blurred boundaries that Starke-Meyerring (2006) refers to may become increasingly clear and crystallized boundaries in the context of online international communication (see also Arewa, 2007; Coombe, 2005).

ACCOUNTABILITY IN INTERNATIONAL CYBERSPACE ENVIRONMENTS

Echoing Huckin (2002), Kathy Bowrey (2006) raises an important issue for technical communication in international online environments: accountability. As its innovators create large digital structures that might be implemented in international contexts, or construct spaces in which users share information (and open themselves up to appropriation), who will be accountable for the end results? In *Law and Internet Cultures*, Bowrey critically deconstructs the law in the context of legal culture, and especially looks at the ways in which U.S. law, practice, and culture has influenced global technology law. While the Internet provides a space for identification, it is not good at providing a space for accountability. One of the problems with international laws is accountability (together with enforcement). Tynes (2007) asserts that because Leonenet cannot punish through sanctioned violence, it cannot be a true state. Its members have no real accountability as they would if they were located in a geography-based state and invested with the power that constitutes a state—the power to discipline. Similarly, the virtual building imagined by architects working together, as described in Haller's (2000) work, has no implications for habitation or human use until it is a physical building located somewhere on earth. Until that time, the virtual building, no matter how real it appears to the architects, remains only a concept in the imagination, or an act of rhetorical invention with no application or embodiment. Whether or not the virtual building will actually deliver on its blueprint promises remains to be seen. Bowrey argues that because responsibility for internet culture and online communicative outcomes has not been assigned, certain structures and "laws" that do not serve socially responsible behavior are perpetuated. Because the historiography (through narratives of the "great" inventors provided via the author-function) of the Internet emphasizes individual genius, responsibility for social outcomes from Internet communication shifts away from larger stakeholders. Huge collaborative projects via the Internet tend to construct what Bowrey describes as the "pseudo-I." Who is responsible for the whole with respect to Web spaces like wikis, social network software, discussion boards, and other online collaboration tools? Bowrey urges that more responsibility should be taken by organizations that have the necessary resources and power. Her argument is relevant for technical communication. As innovators in the field develop technologies for collaboration[9] (and thus appropriation, exploitation, and cross-cultural conflict), methods of accountability should also be developed and shouldered by overseeing organizations that have the necessary resources.

IMPLICATIONS OF CYBERSPACE, INTERNATIONAL LAW, AND RHETORIC FOR TECHNICAL COMMUNICATION

The language and rhetorical turns taken by the judges in the *CCH* opinion show how international differences/divisions are established while leaving space for mutual identification. Originality is something we often take for granted as a required component in the technical writing classroom. Even if students are doing collaborative work, by and large we expect somewhere in their course that students will produce something original. Likewise, in technical communication work environments, one wonders at what point on the *CCH* continuum might a diagram or a set of user instructions become copyrightable due to its originality? Time and labor might count in one jurisdiction but not another. Taking an international view of how originality is defined in the law reveals how a geographical location constructs a definition via a judicial opinion, one that both affirms its independence from other definitions but also admits the influence of other places and ideas (division-identification). Professional writing programmatic concerns might be informed by how various locations differentiate themselves from others in cyberspace. Something as simple as an investigation of originality (or catfish) can illustrate the serious repercussions of defining a single word.

Kenneth Burke's identification and division are two inevitable outcomes in cyberspace, especially when texts-as-intellectual-property circulate globally. Although some have predicted otherwise, it is doubtful that international cyber-environments will cause excessively blurred boundaries or increased homogeneity. Based on rhetorical analyses, division will be as important as ever and may be tied to geographical locations in reaction to the erasure of place in cyberspace. The questions to ask of communication practices in cyberspace are more suited to Foucault's post-author inquiries: What are the modes of existence via rhetoric? Where has it been said? How can it circulate, and what space is there for which subject positions? Technical communicators, innovators, and researchers inventing ways to analyze and explore identification and division in cyberspace will add value to their organizations.

It is interesting to document Peter Yu's (2006) argument on behalf of developing countries. Yu advises less developed countries to exploit the differences between the United States and the EU. St.Amant (2002) documents some of the rhetorical differences between the two entities, but Yu is referring to differences in the valuing of intellectual property protection. Yu argues that developing countries should construct a list of the differing intellectual property protection goals of the United States and the EU and then leverage those differences when negotiating.

I think this idea of exploiting difference/division might prove intriguing for the field of technical communication. Possibilities that might further the investigation of exploiting the affordances of difference could be informed by traditional rhetorical techniques such as *alloiosis* (pointing out difference); *artiistrephon/peristrophe* (turning one's opponent's argument to one's own); *argumentum ex concessis* (reasoning from the premises of one's opponents).[10]

Geographical indicators are a prime example of legally defined modes of pointing out difference. Camembert cheese is asserted to be different from Wisconsin cheese. The difference is supported by international law. For countries that are developing or marginalized in the global economy, one of the smartest moves they might make is to use the arguments of more powerful sovereign entities to their own advantage, as Coombe (2005) points out in her work. If difference is supported in laws such as those that maintain GIs, then first nation and marginalized peoples who formalize the differences of their knowledge by "burnishing images of rustic authenticity onto goods" (Coombe, 2005, p. 12) can leverage the "reasoning of one's opponents" and turn "one's opponent's argument to one's own." In simpler terms, this means seizing the master's tools and using them to one's own advantage. A very recent key example of the implementation of these strategies is that of the enemy combatants at Guantanamo Bay who have used U.S. laws and the U.S. justice system to claim the right to habeas corpus (*Boumediene et al. v. Bush*, 2008). These kinds of strategies map onto the complexities of negotiating Burke's identification and division in cyberspace.

As globalization continues, and as international online spaces multiply in number and in diversity, it is plausible that sovereign entities and/or nation-states will continue to assert their autonomy and will continue to define their boundaries—with the clarity that the Canadian Court provided in the *CCH* opinion. Power will still be a factor in the ways in which geographical locations define their boundaries, and as a rhetorical means of persuasion, one country may adopt the views of another if it needs something (funds, health/medical supplies, military assistance, and so on). St.Amant (2002) writes: "The increased access online communication technologies allow to other cultures does not necessarily mean that cyberspace will necessarily be a multicultural space" (p. 4). He notes that instead, more powerful entities or influences ("cultures") may use online spaces to coerce less powerful cultures into adopting the more powerful countries' rhetorical expectations. That this is an effect of international communication cannot be disputed. For example, Naana E. K. Amonoo (2006), a legal scholar writing as an advocate for intellectual property development in developing countries, especially those in Africa, points out that international exchanges have not created global harmony but instead have exploited less developed countries. To impose patent law in developing countries means that they will have great difficulty adhering to patent law and also tending to medical issues due to the high cost of patented pharmaceuticals. With respect to teaching materials, to impose copyright law in developing countries means that many will go without texts because they will be unable to afford the cost of obtaining proper copyright permission. There have been many instances where pharmaceutical companies have taken local or indigenous knowledge and then patented and exploited it without providing any remuneration to people in the place where this knowledge originated. Kathy Bowrey (2005) provides numerous examples in her book of how U.S. technology law has been ubiquitous on the global scene in determining legal norms and policy.

Basic awareness of international intellectual property laws and their import in cyberspace will benefit technical communicators. While the area is very complicated, a good starting point is the recent Tekom publication (Hennig & Tjarks-Sobhani, 2004) that provides global perspectives from both teachers and practitioners in technical communication—some of the contributors discuss legal issues they face as technical communicators working with conflicting laws and uncertainty in cross-border areas. Other authors who work in this area are Kirk St.Amant (2002) and Rife (2009), as well as numerous legal scholars, such as Kathy Bowrey (2005), Rosemary Coombe (2002, 2005), and Peter Yu (2006).[11]

In technical communication, we need more attention to these issues, more research, from scholars in our field. The area of international intellectual property and its intersection with technical communication is a wide open place. While legal scholars have worked in the area for some time, we have only begun our exploration. Examining and becoming knowledgeable about the influence of international law on communication practices in cyberspace will enable technical communicators to develop strategic thinking and also methods of accountability. Understanding international law in cyberspace may generate knowledge that can be leveraged "into consumer-user value"(Faber & Johnson-Eilola, 2002, p. 139; Reich, 2001, pp. 51–57). "We must find ways to leverage our knowledge and build new knowledge to create and add value in a business culture that is increasingly agnostic to physical products" (Faber & Johnson-Eilola, 2002, p. 141).

Centralizing scholarship on these issues may "enhance the professionalism of the members and the status of the profession" (STC), since our economy is now shaped by high-value modes of production (rather than high-volume modes) (Faber & Johnson-Eilola, 2002; Henry, 2000), and companies will lose competitive edge due to sheer lack of brain power. At the same time, on a global level, technical writers continue to be marginalized within the workplace (Hennig & Tjarks-Sobhani, 2004). Spilka (2002) imagines the field emerging as a profession and seeks methods to "elevate the role of technical communicators so that they become leaders in their organization and the value of their work equals that of others high in the hierarchy of their organizations" (p. 107). Technical writers need to think in terms of return on investment cost, profitability, and ways to gain strategic advantage (Faber & Johnson-Eilola, 2002).

It might also be fruitful to think of cyberspace as a location for rhetorical invention via the imagination—just as Leonenet users imagined and planned for their nation-state to arise once again in a geographical location, and just as architects planning a building discuss the virtual structure as if it already existed. Bowrey (2005) points to this playful aspect of the Internet as well. Drawing upon the works of Lyotard, Deleuze, Guattari, and Appadurai, she describes the unique tale of Mandeville's travels (p. 27). Subsequent to frontier explorations, Pliny described "Blemmyae," men with faces on their chests living in the deserts of Libya. Illustrations of these men were featured in The Travels of Sir John Mandeville, claimed to be the product of an English knight who traveled between 1322 and 1356 (see also Rife, 2006a). Bowrey states: "There is an uneasy connection between the boastfulness of Mandeville's exploits and the unreality of the dot.com bubble, which was accompanied by our own times' equivalent to Mandeville's exciting tales of discovery" (p. 31). Ultimately, Bowrey invites the reader to theorize Internet culture in a way that is not all about lost community, but instead to theorize the Internet through imagination, fantasy, and even the idealized, just as the Leonenet list serve members did. Cyberspace provides a place of sorts to dream up and work out the imagined—but it is always tempered by the law and is eventually accountable to more earthly-material concerns.

This chapter is a small step toward realizing the potential contributions that technical communicators might make via strategic decisions that would impact cyberspace as well as workplace applications, but it also reiterates the importance of rhetoric, rhetorical invention, and the rhetorical tradition, even in cyberspace environments. Rhetoric allows one to "locate, name, analyze, and ultimately influence relations of power that make up our society" (Hart-Davidson, Zappen, & Halloran, 2005, p. 125). Gerard Hauser states that the rhetorical situation is one where interactions between actors create an ongoing struggle between permanence and change, and as such it creates an environment where actors' self-structuring behavior forms "publicness" in an "endless process of negotiation" (quoted in Hart-Davidson et al., 2005, p. 126). Burke's idea of property embodying division in the context of identification and the importance here of rhetoric and the use of rhetorical analysis show that using rhetoric to study cyberspace might be facilitative for purposes of strategizing, planning, even writing rhetorically appropriate texts and using rhetorically appropriate media for those texts in international-cyberspace environments. Such understandings can also lead to public policy arguments (see Porter, 2005, for further discussion of this).

While it might be legal to take innovative ideas, practices, remedies, or products from a developing country, it might not be ethical to do so. Many ancient practices, such as herbal remedies, recipes, even ways of knowing, cannot be protected by any intellectual property law, because, for example, in the United States for an item to be patentable it must be novel, and for an artifact to be copyright protectable it must be original. Therefore, no ancient or traditional practice can be protected, because such practices are neither novel nor original, having been used for generations. So while it might be perfectly legal to exploit the knowledge of marginalized communities, it might be wrong, immoral, unethical, or unjust. These kinds of policy arguments could be developed from within our field. It might be perfectly legal to use someone else's materials in one's own creation, but depending on the context it might not serve justice to do so. On the other hand, while intellectual property law might prohibit the another's materials without express permission, the law might be wrong, unreasonable, out-of-date, and so forth.

In a recent study I conducted, "Is There a Chilling of Digital Communication?" (Rife, 2008), I found that 73% of 384 digital writers, randomly selected technical communication students and teachers in U.S. technical and professional writing programs, indicated in a survey that they approved of following their conscience on exceptional occasions even if it meant breaking the law. In this study, I also interviewed digital writers who indicated that they had used others' materials without permission because they felt it brought no harm, and that it was or should be justified under the U.S. fair use doctrine. In my study, ethics trumped the law. Public policy arguments could be created from within our field arguing both for and against appropriating others' resources, materials, ideas, and so forth, depending on the context.

CONCLUSION

This chapter envisions a world in which technical communicators are not trained merely to support the creation of "physical commodities"(Faber

& Johnson-Eilola, 2002, p. 138). While technical communicators should not practice law or give legal advice, since writing practices are impacted by international intellectual property laws, such communicators as experts in writing should know when legal questions should be asked, when user experiences might invoke law, when formal legal advice might be necessary in order to provide their companies with a strategic advantage. Developing expertise in the area of international intellectual property law, and using rhetoric to do so, sees technical communicators as professionals adding as much value to the organization as any member at any level: workers thinking "beyond product and efficiencies, . . . [engaging] in actual knowledge work. Such work puts technical communicators in the center of business practices" (Faber & Johnson-Eilola, 2002, p. 147).

NOTES

1. I outline a few select international laws in this note:

Berne Convention for the Protection of Literary and Artistic Works (1886): Berne provides minimum protections for copyrighted works at the international level and depends on three basic principles: (1) specifying "national treatment" (defined as treating one's own nationals and foreigners equally with respect to intellectual property protection); (2) specifying no formal registration requirement (in order for the United States to become a member, it had to remove the registration requirement for copyright, which was done through the adoption of the 1976 Copyright Act); (3) specifying that protection for a copyrighted work in a foreign country is independent of protection in the originating country. Some of the exclusive rights protected under Berne include the right to translate, to make adaptations, to perform, recite, communicate, and broadcast in public, and make reproductions. The convention also provides protection for moral rights. Moral rights include the right to attribution, to integrity in a work, and to the publication of unpublished works. While the United States is a signatory to the Berne Convention, it does not provide for moral rights in its copyright statute except under the very narrow provisions of the Visual Artists Rights Act (VARA). Since each country implements the convention and other international standards according to its own model, the U.S. Berne Implementation Act of 1988 interprets the Berne-mandated moral rights narrowly enough for them to be covered by VARA. It is also arguable that the United States provides rights similar to moral rights via other legal protections such as the right to privacy, the right to protection against defamation, the right to publicity, and so on. However, the more economically powerful the country is, the more autonomy it will have in implementing international standards.

In 1886 when Berne was adopted, the world was internationally minded, as a number of other multilateral conventions came into existence as well. Some of these are still in effect today, including the International Telegraph Convention, the Universal Postal Convention, and the Paris Convention for the Protection of Industrial Property (Ricketson, 1986, p. 517). By 1886, most European states had enacted their own copyright laws, and by 1791 the United States and seven states in Latin America had copyright laws as well. Greece, Bulgaria, and Turkey all had authors' rights protected in some form (Ricketson, 1986). However, even with all of the various laws in place, piracy was a normal part of the culture in many locations. The biggest victims were France and the United Kingdom. Irish pirates are credited with pirating works by English authors, while French authors were the specialty of copyright pirates in Belgium (Ricketson, 1986). The Berne Convention tried to address the international piracy problem. An examination of history makes it clear that international standards usually come into existence to address an international problem. Which members of the international community are allowed to define and rhetorically construct international problems is certainly an interesting question, and one that is underresearched in our field. More research on such matters would be welcome.

During the time before the birth of the Berne Convention, several schools of thought existed as to how to provide international protection to copyrighted works. One school of thought recommended that every country should offer equal protection. Ricketson (1986) rightfully describes this as an "idealistic" view, imagining that having universal, identical copyright protection would negate the existence of "time, nationality or territoriality" (p. 522). The convention that was eventually signed did not go very far along the path of universal codification (note that while the United States attended the Berne Conference, its delegate made clear that the United States was participating only as an observer and not as a potential signatory: the United States did not join Berne until 1976). An examination of historical trends shows that when universalization is offered in response to a problem, resistance is encountered on some level.

Madrid System for International Registration of Trademarks: A trademark is a word, symbol, phrase, or design or a combination of those items that identifies the goods of one party and distinguishes them from those of another. Examples include McDonald's, UPS, the Gap, Abercrombie & Fitch, and the logos associated with these. The Madrid System functions under the Madrid Agreement (1892) and the Madrid Protocol (1996) and is administered by the World Intellectual Property Organization (WIPO). The owner of a mark can obtain protection in any member countries of the Madrid Union via a one-time filing in the owner's national or regional trademark office.

Interestingly, while the Madrid Agreement has been in existence for over a century, a number of key countries, including the United States, refused to agree to it. Various reasons have been offered for this refusal. The Madrid System sometimes requires filing in two languages, and even now privileges Spanish and French as primary languages (Ladis & Perry, LLC, 2006). Language choice is always a political issue, and I cannot imagine any time in the near future when the U.S. administration would wish to file any primary international documents, routinely, in a language other than English. The 1996 Madrid Protocol was developed so that the United States would participate in an international trademark protection system. Based on the foregoing, it seems highly unlikely that much of anything, including online communication practices in technical communication, will ever be universalized. Additionally, it appears, from the way international law has developed, that when anything near universalism appears, one or another sovereign entity will refuse to agree, in order to reinscribe its autonomy, power, and so forth.

Agreement on Trade Related Aspects of Intellectual Property Rights (TRIPS): Administered by the WTO, TRIPS is the most comprehensive multilateral intellectual property agreement. TRIPS was negotiated in the Uruguay Round (1986–1994). The outcome of the negotiations was that the General Agreement on Tariffs and Trade (GATT) turned into the World Trade Organization. The negotiations included discussions of issues of international copyright violation and the violation of other forms of intellectual property rights. The Uruguay Round and its outcome have been criticized by some for a lack of attention to the needs of developing countries, along with a focus on the protection of intellectual property rights rather than the protection of the public good or the preservation of fair use. The negotiators were largely concerned with agricultural trade, not with the protection of an international commonality of ideas to benefit humankind (Singh).

TRIPS provides standards showing how member countries might give adequate protection to intellectual property rights, how international intellectual property protection should be applied, and how member countries should enforce rights in their respective jurisdictions. The main provisions of concern to technical communicators are the TRIPS requirements of "national treatment" (treating foreigners equally with one's own nationals); most-favored nation treatment (treating nationals of all WTO member trading partners equally); and the connection of intellectual property protection with support for technological innovation and technology transfer.

European Union Directives: These create standards that are increasingly being incorporated into the technical writing practices of the 27 EU countries. EU Directives set new standards for technical documentation. These standards must be implemented and localized by technical writers preparing documentation for EU countries. However, apart from the EU Directives, each member country also has its own implementation act that interprets and localizes directives in harmony with each country's existing regime. Of course, the United States and other non-EU countries have their own documentation standards and product liability legal regimes.

2. Intellectual property rights are usually defined as rights given to a creator over the creations of her/his mind (World Trade Organization). The rights are given exclusively to the creator but are subject to exceptions and limitations. Intellectual property can be divided into two main areas: copyright and industrial property. Copyright protects literary and artistic works, while industrial property protection includes patent, trademark, and geographical indicator protection. A number of international organizations exist with respect to international trade, but the international organizations particularly focusing on international intellectual property issues are the World Intellectual Property Organization (WIPO) and the World Trade Organization (WTO) (United States Patent and Trademark Office, 2006).
3. See Johnson-Eilola, 1997; Juillet, 2004; Logie, 2005, 2006; Porter, 2005; Selber, 2006; Starke-Meyerring, 2006; Waller, 2006a, 2006b.
4. I discuss the collection fully in Rife, 2007.

5. Moral rights are explained and defined in note 1.
6. In additional to originality, many countries provide that creations must be "fixed" in order to be protected under copyright. This explains the international debate on how to protect traditional knowledge—knowledge that comes from nonfixed oral traditions, and knowledge that is not original because it is often passed down generation to generation. In the United States, in order to receive copyright protection, a creation must be original and fixed.
7. Thatcher does not state where he obtained the population that he surveyed. All the reader knows about the survey is that it included 200 U.S. Americans and 200 South Americans.
8. Thatcher implies that the survey differences within the South American population are due to differences of preference within the population itself. Thus, it is not his purpose to essentialize an entire population.
9. Examples of such developing technologies include resource exchange systems that allow teachers to exchange teaching materials and ideas. I am aware of such databases being developed by William Hart-Davidson of the WIDE Research Center at Michigan State University, but there are several such teacher exchange systems of sorts already in use. For example, Michigan State University has already developed and supports Lon Cappa, which allows teachers to exchange test materials, pictures, images, movies, and so on, through the use of a database to which they contribute. The Society for Technical Communication encourages international exchanges, and some technical communicators have begun list serves to foster such exchanges. I participate with technical communicators from India in a list serve, which is maintained by the STC community in India. The use of open source software and academic journals has received much attention recently—also encouraging digital exchanges at the international level.
10. These definitions are taken from Richard A. Lanham's *A Handlist of Rhetorical Terms*, 2nd edition (1981).
11. Legal scholars have written prolifically on international intellectual property issues, but such work is not localized for our field. For more examples, see Scassa (2004), Tabatabai (2005), Vaver (2004), and Gervais (2004, 2005) All of these authors write on issues of comparison in terms of copyright law, fair dealing, and fair use between Canada and the United States (although they often refer to other countries' laws as well).

REFERENCES

Abdollah, K. et al. (2004). Technical documentation in Finland. In J. Henning & M. Tjarks-Sobhani (Eds.), *Technical communication international: Today and the future* (Vol. 9) (pp. 77–89). (Trans.). Lübeck, Germany: Schmidt-Romhild.

Amonoo, N. E. K. (2006, September 26). Inside views: Should the developing world copy to "catch up" with developed countries? Retrieved September 27, 2006, from the Intellectual Property Rights Web site: http://www.ip-watch.org/weblog/index.php?p=406&res=1280&print=0

Arewa, O. B. (2007). Culture as property: Intellectual property, local norms and global rights. Retrieved May 16, 2007, from http://works.bepress.com/o_arewa/9/

Bazerman, C. (1999). *The languages of Edison's light.* Cambridge, MA: MIT Press.

Berne Convention for the Protection of Literary and Artistic Works. (1886). Retrieved September 30, 2006, from the WIPO database of intellectual property, http://www.wipo.org/treaties/en/ip/berne/pdf/trtdocs_wo001.pdf

Berne Convention Implementation Act of 1988, 100th Cong., 2nd Sess. (1988). Retrieved September 30, 2006, from http://www.copyright.gov/title17/appendixii.pdf

Block, D. (2004). Globalization, transnational communication and the Internet. *International Journal on Multicultural Societies, 6*(1), 13–28.

Boumediene et al. v. Bush, President of the United States et al., 553 U.S. (2008).

Bowrey, K. (2005). *Law and Internet cultures.* Cambridge, MA: Cambridge University Press.

Brockmann, R. J. (1988). Does Clio have a place in technical writing? *Journal of Technical Writing and Communication, 18,* 297–304.

Brockmann, R. J. (1998). *From millwrights to shipwrights to the twenty-first century: Explorations in a history of technical communication in the United States.* Cresskill, NJ: Hampton.

Brockmann, R. J. (1999). Oliver Evans and his antebellum wrestling with theoretical arrangement. In T. C. Kynell & M. G. Moran (Eds.), *Three keys to the past: The history of technical communication* (pp. 63–89). Stamford, CT: Ablex.

Bureau of Public Information. (2005, October 20). General conference adopts convention on the protection and promotion of the diversity of cultural expressions. Retrieved September 30, 2006, from the United Nations Educational, Scientific and Cultural Organization Web site: http://portal.unesco.org/culture/en/ev.php-URL ID=29078&URL_DO=DO_TOPIC&URL_SECTION=201.html

Burke, K. (1969). *A rhetoric of motives.* New York: Prentice-Hall.

Catfish Farmers of America. (2005). *Catfish Farmers of America Convention*. Retrieved May 15, 2007, from http://www.catfishfarmersamerica.org/convention.html

CCH Canadian Ltd. v. Law Society of Upper Canada, SCC 13, 30 C.P.R. (4th) 1, (2004).

Cloatre, E. (2008). Socio-TRIPS and pharmaceutical patents in Djbouti: An ANT analysis of socio-legal objects. *Social Legal Studies, 17,* 263–281.

Comaroff, J., & Comaroff, J. (Eds.). (1993). *Modernity and its malcontents: Ritual and power in postcolonial Africa.* Chicago: University of Chicago Press.

Coombe, R. J. (2002). The recognition of indigenous peoples' and community traditional knowledge in the international law. *St. Thomas Law Review, 14*(2), 275–285.

Coombe, R. J. (2005). Legal claims to culture in and against the market: Neoliberalism and the global proliferation of meaningful difference. *Law, Culture and Humanities 1*(1), 32–55.

Copyright Act of Canada. Retrieved April 30, 2006, from http://www.cb-cda.gc.ca/info/act-e.html

DeVoss, D. N., & Porter, J. E. (2006). Why Napster matters to writing: Filesharing as a new ethic of digital delivery. *Computers and Composition, 23*(2), 178–210.

Durack, K. T. (1997). Gender, technology, and the history of technical communication. *Technical Communication Quarterly, 6*, 249–260.

Durack, K. T. (2001). Research opportunities in the US patent record. *Journal of Business and Technical Communication, 15*(4), 490–510.

Durack, K. T. (2006). Technology transfer and patents: Implications for the production of scientific knowledge. *Technical Communication Quarterly, 15*(3), 315–328.

Faber, B., & Johnson-Eilola, J. (2002). Migrations: Strategic thinking about the future(s) of technical communication. In B. Mirel & R. Spilka (Eds.), *Reshaping technical communication: New directions and challenges for the 21st century* (pp. 135–148). Mahwah, NJ: Erlbaum.

Feist Publications, Inc. v. Rural Telephone Service Co., 499 U.S. 340 (1991).

Fishman, T. C. (2005, January 9). Manufaketure. *New York Times Magazine,* 40.

Foucault, M. (1984). What is an author? In P. Rainbow (Ed.), *The Foucault reader* (pp. 101–120). New York: Pantheon.

Gervais, D. J. (2004). Canadian copyright law post-CCH. *Intellectual Property Journal, 18,* 131–167.

Gervais, D. J. (2005). The purpose of copyright law in Canada. *University of Ottawa Law and Technology Journal, 2*(2), 316–356.

Giliarevski, R. S. (2004). Technical documentation in Russia. In J. Henning & M. Tjarks-Sobhani (Eds.), *Technical communication international: Today and the future* (Vol. 9, pp. 129–148). (Trans.). Lübeck, Germany: Schmidt-Romhild.

Gilliam v. American Broadcasting Companies, Inc., 538 F. 2d 14 (2nd Cir. 1976).

Gurak, L. J. (2005). Ethics and technical communication in a digital age. In M. Day & C. Lipson (Eds.), *Technical communication and the World Wide Web* (pp. 209–221). Mahwah, NJ: Erlbaum.

Halavais, A. (2000). National borders on the World Wide Web. *New Media and Society, 1*(3), 7–28.

Haller, C. R. (2000). Rhetorical invention in design: Constructing a system and spec. *Writing Communication, 17*(3), 353–389.

Hart-Davidson, W., Zappen, J. P., & Halloran, M. S. (2005). On the formation of democratic citizens: Rethinking the rhetorical traditional in a digital age. In R. Graff, A. E. Walser, & J. M. Atwill (Eds.), *The viability of the rhetorical tradition* (pp. 125–140). Albany, NY: State University of New York Press.

Hennig, J., & Tjarks-Sobhani, M. (2004). *Technical communication international: Today and the future* (Vol. 9, Trans.). Lübeck, Germany: Schmidt-Romhild.

Henry, J. (2000). *Writing workplace cultures: An archaeology of professional writing.* Carbondale: Southern Illinois University Press.

Herrington, T. K. (1997, Spring). The unseen "other" of intellectual property law or intellectual property is not property: Debunking the myths of IP law. *Kairos 3*(1).

Herrington, T. K. (1998). The interdependency of fair use and the first amendment. *Computers and Composition, 15*(2), 125–143.

Herrington, T. K. (2001). *Controlling voices: Intellectual property, humanistic studies, and the Internet.* Carbondale: Southern Illinois University Press.

Herrington, T. K. (2003). *A legal primer for the digital age.* New York: Pearson Longman.

Howard, T. W. (2004). Who "owns" electronic texts? In J. Johnson-Eilola & S. A. Selber (Eds.), *Central works in technical communication* (pp. 397–406). New York: Oxford University Press.

Huckin, T. (2002). Globalization and critical consciousness in technical and professional communication. Paper presented at the Conference of the Council of Programs in Scientific and Technical Communication. Retrieved April 30, 2006, from http://www.hum.utah.edu/uwp/globalization/

Jessop, B. (1999). Reflections on globalization and its (il)logic(s). Lancaster, England: University of Lancaster. Retrieved October 4, 2006, from http://www.lancs.ac.uk/fass/sociology/papers/jessop-reflections-on-globalization.pdf

Johnson-Eilola, J. (1997, Spring). Intellectual property: Questions and answers. *Kairos, 3*(1).

Juillet, C. (2004). Protect your Web site from legal land mines. *Intercom.* Retrieved November 23, 2006, from http://www.stc.org/intercom/PDFs/2004/2004011_06-07.pdf

Ladas & Perry, LLP. (2004, March). Madrid system—Changes. Retrieved September 28, 2006, from http://www.ladas.com/BULLETINS/2004/0304Bulletin/Madrid_Changes.html

Lam, T. D. T. (2003). US "catfish war" defeat stings Vietnam. *Asian Times Online.* Retrieved May 16, 2007, from http://www.atimes.com/atimes/Southeast_Asia/EG31Ae02.html

Lanham, R. A. (1991). *A handlist of rhetorical terms* (2nd ed.). Berkeley: University of California Press.

Logie, J. (2005). Parsing codes: Intellectual property, technical communication, and the World Wide Web. In M. Day & C. Lipson (Eds.), *Technical communication and the World Wide Web* (pp. 223–241). Mahwah, NJ: Erlbaum.

Logie, J. (2006). *Peers, pirates, and persuasion: Rhetoric in the peer-to-peer debates.* West Lafayette, IN: Parlor Press.

Migdal, J. (2001). *State in society: Studying how states and societies transform and constitute one another.* Cambridge, MA: Cambridge University Press.

Porter, J. E. (2005). The chilling of digital information: Technical communicators as public advocates. In M. Day & C. Lipson (Eds.), *Technical communication and the World Wide Web* (pp. 243–259). Mahwah, NJ: Erlbaum.

Porter, J. E., & Rife, M. C. (2005). MGM v. Grokster: Implications for educators and writing teachers. Retrieved November 18, 2006, from http://www.ncte.org/cccc/gov/committees/ip/125704.htm

Reich, R. B. (2001). *The work of nations: Preparing ourselves for 21st century capitalism.* New York: Alfred A. Knopf.

Ricketson, S. (1986). The birth of the Berne Union. *Columbia VLA Journal of Law and Arts, 11*, 517–526.

Rife, M. C. (2007). Book review [Review of the book *Technical communication international: Today and the future: Vol. 9. Schriften zur Techische Kommunikation, Band 8*]. *Journal of Business and Technical Communication, 21*, 227–232.

Rife, M. C. (2009). Fair dealing in Canada, fair use in the US: Network innovation in the judicial opinion. In *Writing in the Knowledge Society.* D. Starke-Meyerring, A. Paré, N. Artemeva, M. Horne, & L. Yousoubova (Eds.). West Lafayette, IN: Parlor Press.

Rife, M. C. (2006a). H-NET Book Review [Review of the book *Law and Internet cultures*]. Retrieved May 21, 2007, from http://www.h-net.org/reviews/showrev.cgi?path=175901160495633

Rife, M. C. (2006b). Local global literacies for judges: Canadian/US judicial opinion approaches to globalization in copyright contexts. Retrieved September 23, 2006, from http://papers.ssrn.com/sol3/papers.cfm?abstract_id=900743

Rife, M. C. (2008). Is there a chilling of digital communication? Exploring how knowledge and understanding of fair use influence Web composing. Doctoral dissertation, Michigan State University. In progress.

Ring, P. (2004). Technical documentation in Denmark. In J. Hennig & M. Tjarks-Sobhani (Eds.), *Technical communication international: Today and the future* (Vol. 9, pp. 51–64, Trans.). Lübeck, Germany: Schmidt-Romhild.

Ross, M. (2002). Ross applauds catfish farmers of America for taking on Vietnamese imports. Retrieved May 15, 2007, from http://www.house.gov/ross/pr_2001_2002/pr_062802b.html

Scassa, T. (2004). Recalibrating copyright law? A comment on the Supreme Court of Canada's Decision in *CCH Canadian Limited et al. v. Law Society of Upper Canada. Canadian Journal of Law and Technology, 3,* 89–100.

Selber, S. (2006). Beyond Napster: Institutional policies and digital economics. Retrieved June 25, 2006, from http://www.wide.msu.edu/widepapers/Selber_responsetempl2.pdf/file_view

Singh, S. (n.d.). Uruguay round—A historical perspective. Retrieved September 23, 2006, from the Third World Network Web site: http://www.twnside.org.sg/title/hist-cn.htm

Society for Technical Communication. (2006). Retrieved August 25, 2006, from http://www.stc.org

Spilka, R. (2002). Becoming a profession. In B. Mirel & R. Spilka (Eds.), *Reshaping technical communication: New directions and challenges for the 21st century* (pp. 91–96). Mahwah, NJ: Erlbaum.

St.Amant, K. (2002). *Culture, conflict, and cyberspace: A case study of European Union-United States negotiations over the European Union Data Protection Directive and the United States Safe Harbor Principles.* Unpublished doctoral dissertation, University of Minnesota.

Starke-Meyerring, D. (2005). Meeting the challenges of globalization: A framework for cultural literacies in professional communication programs. *Journal of Business and Technical Communication, 19*(4), 468–499.

Sun, H. (2004). *Expanding the scope of localization: A cultural usability perspective on mobile text messaging use in American and Chinese contexts.* Unpublished doctoral dissertation, Rensselaer Polytechnic Institute.

Tabatabai, F. (2005). A tale of two countries: Canada's response to the peer-to-peer crisis and what it means for the United States. *Fordham Law Review, 73*, 2321–2381.

Thatcher, B. (2001). Issues of validity in intercultural professional communication research. *Journal of Business and Technical Communication, 15*(4), 458–489.

Trade Related Intellectual Property Rights Agreement (TRIPS). Retrieved September 30, 2006, from http://www.wto.org/english/docs_e/legal_e/legal_e.htm#TRIPs

Turner Entertainment Co. v. Huston, CA Versailles, civ. ch. (1994). Translated in Ent. L. Rep., Mar. 1995, at 3.

Tynes, R. (2007). Nation-building and the diaspora on Leonenet: A case of Sierra Leone in cyberspace. *New Media and Society, 9*, 497–518.

UNESCO. (2001). Universal Declaration of Cultural Diversity. Retrieved September 30, 2006, from http://unesdoc.unesco.org/images/0012/001271/127160m.pdf

United States Patent and Trademark Office. (2006). Retrieved September 23, 2006, from http://www.uspto.gov/main/profiles/international.htm

United States Patent and Trademark Office. (2007). Geographical indication protection in the United States. Retrieved May 16, 2007, from http://www.uspto.gov/web/offices/dcom/olia/globalip/pdf/gi_system.pdf

Vaver, D. (2004). Canada's intellectual property framework: A comparative overview. *Intellectual Property Journal, 17,* 125–180.

Verhein-Jarren, A. (2004). Technical documentation in Switzerland. In J. Hennig & M. Tjarks-Sobhani (Eds.), *Technical communication international: Today and the future* (Vol. 9, pp. 25–35, Trans.). Lübeck, Germany: Schmidt-Romhild.

Visual Artists Rights Act of 1990 (VARA). 17 USC, S. 106A, 101st Cong. (1990).

Voichita-Alexandra, G. (2004). Technical documentation in Romania. In J. Hennig & M. Tjarks-Sobhani (Eds.), *Technical communication international: Today and the future* (Vol. 9, pp. 106–115, Trans.). Lubeck, Germany: Schmidt-Romhild.

Waller, R. D. (2006a). Fair use. Ethics case. *Intercom.* Retrieved November 23, 2006, from the Society for Technical Communication Web site: http://www.stc.org/intercom/PDFs/2006/20064_40-41.pdf

Waller, R. D. (2006b). Responses to "fair use." Ethics case. *Intercom.* In J. G. Bryans, Column Ed.). Retrieved November 23, 2006, from the Society for Technical Communication Web site: http://www.stc.org/intercom/PDFs/2006/20069-10_39.pdf

Weber, M. (1947). *The theory of social and economic organization.* New York: Free Press.

World Intellectual Property Organization (WIPO). (2006). Retrieved October 4, 2006, from http://www.wipo.int/portal/index.html.en

World Trade Organization (WTO). (2006). WTO | Welcome to the WTO Web site. Retrieved September 23, 2006, from http://www.wto.org/

Yu, P. K. (2006). TRIPS and its discontents. *Marquette Intellectual Property Law Review, 10*, 369.

SECTION II

Online Interactions Between Cultures

CHAPTER 5

Pleasure in Naming All the Parts of the Known in Their Expected Order: How Traditional Chinese Agrarian Culture Influences Modern Chinese Cyberspace Communication

Daniel D. Ding

CHAPTER OVERVIEW

The Confucian concept of *naming* helps us uncover patterns in Chinese Web design, especially in displays of verbal information. In this chapter, I examine the influence of the practice of "naming the known" on Chinese Web design by focusing in particular on the Web site of the Chinese Foreign Ministry (CFM). To help us see the unique features of the CFM Web site, I contrast it with the Web site of the US Department of State (USDS). My discussion suggests that, while this strategy of naming the constituents is not limited to online documents, it serves two rhetorical purposes: to ensure the continuity of the entire procedure and to filter out unnecessary information.

Since the Internet was introduced to China in the mid-1990s, its use has been growing rapidly. The number of Internet users doubled every year in the past 10 years, and by the end of 2007, China had about 140 million Internet users (Li, 2007). In fact, some experts predict that the number of Chinese Internet users will soon outnumber those in the United States (Li, 2007). This rapid development of Internet use has greatly boosted the Chinese economy. Numerous businesses sell their products and place advertisements on the Internet so that

Chinese consumers can access products more quickly and easily than 20 years ago. Internet cafés are flourishing in small towns as well as in major cities. Web-related advertisements are plastered on buses, taxis, trains, and walls, while Web-related billboards are to be seen everywhere. Even the Communist Party of China (CPC) Central Committee encourages online interactions, claiming that the Internet is a "knowledge economy" (Hachigian, 2001, p. 119). With the help of the CPC Central Committee, every ministry in the Chinese government now has its own Web site.

Widespread Internet access in China also facilitates international exchanges between China and the rest of the world. Chinese enterprises have developed their international potential and are expanding into overseas markets. This development means that more U.S. enterprises and organizations will engage in business interactions with Chinese companies. Online activities play an important role in these interactions and in a growing number of business transactions.

Online communication is a major bridge between Chinese and U.S. enterprises. Such interactions, however, are also bound to create problems for Web users, as some researchers have already pointed out (McCool, 2006; St.Amant, 2005, 2007). St.Amant in particular suggests that different cultural expectations of Web design can lead to "mismatched cultural expectations [that result] in miscommunication related to visual displays" (2005, p. 76). Therefore, the bridge builders—professional communicators—need, among other things, to understand the cultural expectations that underlie Chinese Web practices.

To understand how Chinese cultural expectations affect Chinese Web design, I propose that the Confucian concept of *naming* can help individuals uncover a pattern in Chinese Web design—especially in the display of verbal information. In this chapter, I examine the influence of the Confucian practice of "naming the known" on Chinese Web design by focusing on the Web site of the Chinese Foreign Ministry (CFM). To help readers see the unique features of the CFM Web site, I will contrast this Chinese site with the Web site of the U.S. Department of State (USDS). First, I introduce the Chinese cultural practice of naming the known. Next, I compare and contrast the two Web sites and focus on the different types of information found on these sites. Finally, I examine the significance—and the rhetorical purpose—of naming the known in Chinese professional communication.

NAMING THE KNOWN: A MAJOR FEATURE OF CHINESE CULTURE

Confucianism, a philosophy that has influenced Chinese society for more than 2,500 years, is actually centered around agriculture. It is the "theoretical expression of different aspects of the life of the farmers" (Fung, 1997, p. 20). So, as Fung has pointed out, Confucianism often expresses what farmers feel but are not capable of expressing themselves because they have received very little education. The major purpose of Confucianism is to justify and theorize the farmer-based social system of Confucius's time.

Confucius's Advocacy of the Use of Correct Names

The philosopher Confucius was mainly concerned with the social order and stability of the farmer-based society in which he lived. He firmly believed that the only way to maintain social stability and keep order in such a society was to arrange affairs according to actuality, and "the actual must in each case be made to correspond to the name" (Fung, 2003, p. 59). To achieve the objective, correct names must be used, and incorrect names must be changed or rectified. Although Confucius was mainly concerned with the names of individuals, which indicate family and social relations, he also paid attention to names of objects (Gao & Handley-Schachler, 2003; Hagemann, 1986; Herr, 2003). For example, "When the gentleman names something, the name is sure to be usable . . . and practicable" (Confucius, 1979, p. 118). Being "practicable," for Confucius, means that the names can be used to identify the essence of the objects which bear those names. To illustrate his point, Confucius used the word *gu*—which means "a cornered drinking cup" in Chinese. In *The Analects*, when seeing a drinking cup that does not have corners but is still called a *gu*, Confucius (1979) exclaims that a new name must created for that cup (p. 84). That is, when the corners are not present, the cup must be identified as something other than a *gu*. In such cases of misnaming, the incorrect name must be corrected so that social order and stability can be maintained.

Naming, Identity, and Social Order

In correcting names, Confucius emphasizes an object is named or identified based on the "essence"—or the key characteristics—of the object. Names, in this respect, define the essence of things. When the name of an object matches the essence of that object, the name matches reality and is correct and credible.

From the Confucian perspective, using correct names ensures clarity, avoids ambiguity, and promotes social harmony in society. *Yi Jing*, one of the six Confucian classics, suggests that when correct names are used, "all under Heaven will be established" (Legge, 1993, p. 242). The idea is that human society and nature repeat themselves in cycles just as farmers follow the changing seasons. In order for human society to be stable and proper, people must follow these cycles. In Fung's (1997) words, people must "acknowledge the inevitability of the world as it exists. . . . If we can act this way, we can, in a sense, never fail" (p. 45). The correct use of names, in turn, facilitates the smooth transition from cycle to cycle and maintains social harmony and social order.

Small cycles are manifest in the daily social activities of all people and are, in Owen's terms, "constituents" of large cycles (2000, p. 291). For a cycle to be successfully competed, each participant must perform his or her correct role and actions, and each participant's activities are related to his or her status, position, or name and title. Thus, someone named as a farmer must perform certain tasks at certain times for a cycle to be correctly completed and for harmony and order to be maintained.

To ensure the completeness of large cycles, all individuals must first ensure the completeness of small cycles. To ensure the completeness of small cycles, people must correctly name the constituents of the small cycles. Within this process, the use of correct names helps clarify individual roles and responsibilities, human activities, and recipients of human actions—the constituents of the repeating cycles of social processes. When correct names are used, individuals can perform their roles by conforming to the essence defined by their names. In this way, correct naming serves as a foundation for social harmony and social order.

Naming in a Web-Based Context

Chinese professional communicators, including Web designers, use correct names to avoid ambiguity in communication processes. Often, they feel compelled to clarify even a known term, so a document mainly consists of reproductions of categories of things. For example, the Beijing Olympic Organization Committee Web site (BOOC, 2008) provides instructions to help ticket-buyers pay for the tickets they wish to purchase. The instructions consist of three options: first, "Pay with Visa Card"; second, "Pay with Bank of China Savings Account"; and third, "Pay with Cash" (BOOC, 2008). The first option defines "Visa Card" by using pictures of four Visa cards issued by different banks. The second option defines a savings account and suggests that buyers use their own local banks. The third option merely tells buyers to go to a designated bank.

The instructions for making payments do not specify the actions that buyers need to perform in order to accomplish the task of paying for a ticket. (For instance, can they pay over the phone? If yes, what number and time should they call?) Instead, the instructions consist of the names of the various things needed for the process of making payments. It seems that, Chinese professional communicators believe that, if names are defined correctly, the user will be able to intuit what he or she needs to do in order to finish the overall task (see also Wang, 2000). This approach implies that readers who are unfamiliar with a specific type of process will have some difficulty determining what to do.

This practice of using correct names to ensure the completeness of the cycle of making payments seems to corroborate Hall's (1997) and Stewart and Bennett's (1991) notions of high-context culture. In such cultures, individuals' thinking style "involves a high degree of sensitivity to context [and] relationships [among individual events]" (Stewart & Bennett, 1991, p. 42). High-context cultures also stress the unity between the context in which interactions take place and the individual events and objects present in that context. To put it in a nutshell, in "high-context communication, most of the information is either in the physical context or internalized in the person" (Hall, 1997, p. 47). According to this perspective, correct names could help an audience determine what actions its members should take once they have the components of the overall context before them (see also Ding, 2003; Wang, 2000). Names help individuals identify the context in which they are interacting, and that context provides a series of implied rules on how to communicate in that situation.

She Jing (Book of Songs): Proponent of Naming Things

The *She Jing*, the first Chinese collection of songs, is perhaps the preeminent classic out of all the six Confucian classics. It exemplifies many Confucian thoughts, especially the Confucian concept of naming things. As Owen (1996) has rightly pointed out, the *She Jing* was central to Confucian educational philosophy because, through naming objects and human activities, it helped individuals respond properly and naturally "to the events of life" (p. xiv). These "events" are agrarian events and activities: four seasons, human rituals and rites, men and women's farming activities, activities including animals, seeds, plants, crops, farming tools, and so forth. One way in which the *She Jing* helps individuals respond to these events and activities is naming their "constituents." By identifying these constituents, people would learn the names of the constituents/parts—and the human actions and recipients of human actions related to such constituents/parts—and could perform prescribed actions with the correct items in order to help complete agrarian tasks effectively. Actually, in the *Shi Jing*, readers always find the authors naming farmers, their farming activities, animals, plants, and crops. For example, Poem 290 narrates farmers' activities related to grass (italics added):

> *Mowing grass, felling trees*
> *Ploughs opening up* the *ground.*
> *A thousand pair* tug at *weeds* and *roots*,
> Along the *paddies*, along the *ridges*.
> Here the *master*, here the *eldest son*,
> There the *brothers*, there the *clan*.
> Here the *yeoman*, here the *hands*,
> Hungry for their field's *meals*,
> *Dainty wives* and *sturdy men.*

Poem 154, the Seventh Month, simply repeats all the natural events that will occur in a year, such as the coming of frost and wind. In these and other poems in the *Shi Jing*, the authors simply accumulate the names of events and actions that are expected to happen in agrarian processes. Such naming, in turn, seems to help guarantee that farmers will be able to complete their work successfully. The significance of accumulating names associated with agriculture lies in the fact that these names represent the completion of a cycle.

As noted earlier, Confucius believed that using correct names helps human society and nature to repeat themselves harmoniously. Farmers always want an agrarian cycle to proceed as has been predicted; that is, all events and human activities should occur in the expected order to guarantee a bumper harvest. With a bumper harvest, society will remain stable and harmonious. Thus, correct naming helps people to perform the correct actions needed to guarantee a successful harvest, and such harvests help ensure social stability. As a result, naming equals predictability, and predictability allows for success and harmony. In Poem 154, farmers use the listing of a collection of names to reveal their desire to complete an agrarian cycle uninterrupted by unexpected events.

It is precisely this feature of naming things in the *Shi Jing* that appealed to Confucius, who claimed that the poems in the *Shi Jing* "stimulate the mind, can train the observation . . . and can alleviate the vexation of life. And in these [poems] one may become widely acquainted with the names of birds, beasts, plants and trees" (Confucius, 1979, p. 145). It is perhaps for this reason that the *Shi Jing* is often ranked first among the Confucian classics.

Naming and Web Site Design

The traditional practice of naming things may enable us to perceive previously unnoticed patterns in the visual displays used to present information on a Chinese Web site. For example, the Chinese Ministry of Science and Technology (MOST), on its Web site (www.most.gov.cn/eng), has issued a call for papers for the Fourth International Symposium on Soft Science (MOST, 2007). The announcement, however, never specifies the procedure for submitting a paper, though it does announce the deadline for such submissions. Instead, the site lists all the topics and themes the symposium will cover, as well the as names of all the organizers, including the chair, deputy chairs, secretary-general, advisors, and others. The site even lists the names of all the speakers at the previous three symposia. (Perhaps naming the speakers at previous symposia is intended to encourage individuals to research them in order to identify previously successful models that can be used to draft successful new papers for the new event.)

This Chinese tradition of naming things may also be responsible for the large amount of information that "Chinese webmasters cram into a single page" (Rogers, 2008, p. 11). After examining five Chinese Web sites that each contain large amounts of information on a single page, Rogers (2008) advises international Web designers to "make as much important information available as possible" when they develop online pages for Chinese users (p. 13).

THE CFM WEB SITE: AN EXAMPLE OF NAMING

In this section, I examine the official Web site of the Chinese Foreign Ministry (CFM). This analysis focuses on the CFM's visa application procedure page (CFM, 2002)—particularly on how the page's features (naming the constituents of a cycle of visa application, such as the various categories of visas) are designed to help Web users go through the entire procedure of applying for a Chinese visa. I also compare and contrast the CFM Web site with that of the U.S. Department of State (USDS), which provides parallel information on how to apply for a U.S. visa.

I chose the CFM Web site for two major reasons:

- It is a very popular Chinese Web site. A growing number of individuals (over 100 million per year) visit China for business, tourism, and other purposes (NTAC, 2006). In addition, the Web site releases weekly updates about China to news agencies all over the world. Thus, both international visitors and news reporters visit the Web site regularly.
- The site has a corresponding U.S. Web site that focuses on similar products and procedures. As a result, it is possible to compare and contrast the two sites to highlight the unique features of the CFM Web site.

The splash page of the CFM Web site contains many links—one of which is "Travel to China." Under this link is a short "Kind Reminder" statement that reads as follows:

> You are warmly welcome to China! To make your trip to China an easy one and pleasant one, you are kindly advised to consult the local Chinese Embassies or Consulate-General first for travel information. They will help you go through the necessary procedures.
>
> We also hope that you find the following "Visa Information" and "Useful Links" helpful (CFM, 2006, para. 6).

At this point, Web users may want to open "Useful Links," believing that this will take them to a local Chinese embassy or a consulate general. But instead, "Useful Links" connects users to various Chinese travel agencies and to the Information Office of the State Council of China http://www.scio.gov.cn). None of these agencies is authorized to issue Chinese visas.

If Web users select the "Visa Information" link, they access a page containing three links:

- Visa Application Form of the PRC
- A Brief Introduction to Chinese Visa
- A Brief Introduction to Visa Application Procedure

If Web users open the third link ("A Brief Introduction to Visa Application Procedure"), they will be led to a page containing a general definition of a Chinese visa, followed by definitions of each of the eight visa categories. Only at the end of each definition can users find a few sentences that describe the documents needed for visa applications. Such descriptions, moreover, are quite brief. For example,

> Visa F: issued to an applicant who is invited to China on a visit, or a study or lecture, business tour, for scientific-technological and cultural exchanges, for short-term refresher course or for job-training, for a period of no more than six months. To apply for a Visa F, the invitation letter from the inviting unit or the unit notification letter/telegram from the authorized unit is required. (CFM, 2006)

At the end of the list of the nine visa categories, applicants are reminded that other documents, such as a valid passport and a completed form, as well as a visa fee, are also required to obtain a Chinese visa. The page concludes with a definition of the "Overseas Chinese Visa Authorities." (The entire page, "A Brief Introduction to Chinese Visa and the Procedure for Visa Application," is included in Appendix 1.) The Web design factors described above reflect Confucian perspectives on naming as related to performing an action correctly.

Confucianism, Naming, and Web Design

Confucianism holds that the *Shi Jing*, with its style of naming constituents, can train individuals to use language well. This includes teaching individuals to repeat names and categories in order to communicate the meanings of objects and items (Confucius, 1979). On the CFM Web site, Chinese professional communicators seem to practice this Confucian approach to language by naming the constituents of the visa application procedure. The idea is that naming these constituent parts helps Web designers ensure the continuity of the application procedure by selecting information relevant to all Web users. In this way, Chinese Web designers appear to communicate their meanings clearly to users—namely, "I want you to learn the names and know what they mean before you perform the task." Once they know these names, users can then perform the related task correctly.

Even when they do provide a step-by-step procedure, as is seen on the Web site of the Chinese Embassy in Washington, DC, Chinese Web designers do not describe how to perform a task. Rather, they show users the positions of the steps in the procedure in relation to each other. In other words, the designers just name these steps and assume that once they are identified, the user will have all that is needed to intuit how to perform each step correctly. For example, Step 1 tells visa applicants to pick a number and sit down to be called (Chinese Embassy, 2006).

The site does not tell users where to obtain that number, nor does it explain how long applicants need to wait to be called, or whether applicants can use the same number at a different time if they are not called on the day they originally selected the number. This example reveals the high-context preference for letting individuals intuit how to behave based upon the context in which the information is presented. Thus, even when specifying steps in a procedure, Chinese Web designers only name the steps in a process and do not provide specific explanations of how to perform those steps.

In essence, the information Chinese visa applicants need is in this list of steps. The information, however, looks more like an inventory list, naming items such as visa categories and required documents, than a set of instructions or a process for applying for a visa. In fact, it seems to provide visa applicants with more information on Chinese visa types than on the actual application procedure. Notably, the parallel USDS Web site takes a very different approach to achieving a similar objective.

THE USDS WEB SITE

The U.S. Department of State (USDS) Web site seems more task oriented than its Chinese counterpart. The link "Travel and Business" on the splash page of the U.S. site takes users to a "Visas" page. Users may then select from two options: "Visas for Americans" or "Visas for Foreign Citizens." The latter link leads U.S. visa applicants to the "Destination USA" page, where users will find general information about U.S. visa policy and procedure. Six links on that page help non-U.S. citizens locate visa information specific to their countries. One of these links is "Locate a Consulate Office," which takes users to a page that contains an alphabetically organized list of every country and city in the world that has a U.S. consulate office (USDS, 2006). Figure 1 shows a screen shot of this page.

For the purposes of this chapter, I chose the city of Beijing, and was connected to the Web page of the U.S. Embassy in Beijing. From there, the hyperlink "Visa to the U.S." took me to the U.S. visa information site, which provides information about U.S. nonimmigrant visas, visa appointments, appointment waiting time, and visa processing time. This site also provides individuals with a useful link entitled "How to Apply for a Visa." This link accesses a page that provides a step-by-step procedure for completing the task of applying for a visa. The procedure consists of three general steps:

- Steps in Making an Appointment
- Complete Visa Forms
- On the Day of Interview

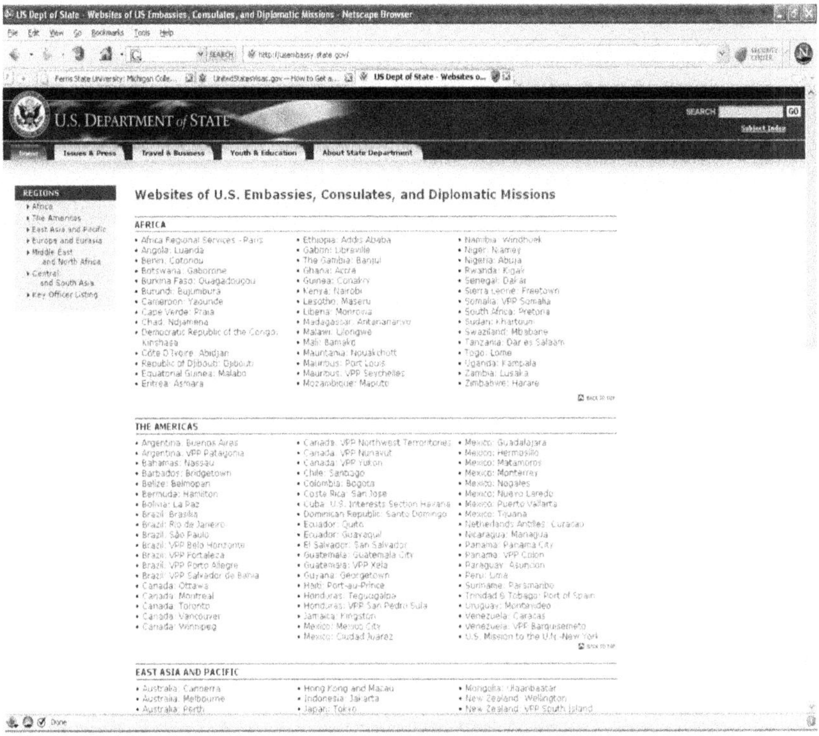

Figure 1. Screen shot of USSD "Locate a Consulate Office" page.
Source: http://usembassy.state.gov

Each general step consists of three to four substeps. For example, Step 1 contains detailed information about setting up a visa appointment:

- Steps in Making an Appointment
  a. Buy a pre-paid PIN card from China CITIC Bank or visit the Visa Information Call Center website to purchase a PIN number online. PIN cards cost 54 RMB for 12 minutes of phone time or 36 RMB for 8 minutes. Any unused minutes can be used at a later time or transferred to another party.
  b. You may also pay the visa application fee (RMB 810) at China CITIC Bank. Please note that this fee is non-refundable and the original receipts are valid for one year.
  c. Call the Visa Information Call Center to make an interview appointment or ask specific questions about the visa process. Please ensure that you

know the following information: applicant's full name, passport number, ID number, contact information, purpose of travel, place of residence, whether s/he was refused before, etc. *All long distance charges are the sole responsibility of the caller, and all calls are generally answered in less than 30 seconds.*

Call Center Numbers:
Within China: 4008-872-333 (toll free)
(021) 3881-4611 (Shanghai local number)
Outside China: (86-21) 3881-4611
Hours: Monday–Friday from 7:00 a.m. p 7:00 p.m.
Saturday from 8:00 a.m.–5:00 p.m.
Pre-recorded visa information is available 24 hours a day, 7 days a week (U.S. Embassy, 2006)

(See Appendix 2 for the complete page, "How to Apply for a Visa.")

The instructions provided on the USDS Web site are task- or procedure-based in contrast to the listing approach used on the CFM website. The USDS instructions tell users what to do in order to set up a visa appointment and even provide information on how to purchase a phone card and how to pay the fee.

The Chinese Web site establishes visa categories, while the .US. Web site presents steps for completing the visa application procedure. This reveals how the Confucian preference for naming influences Web site design in Chinese culture.

TWO RHETORICAL PURPOSES SERVED BY NAMING PARTS

As the CMF Web site illustrates, the Chinese visa application procedure is represented by defining—or identifying—the constituents/parts of the procedure. Naming these constituents, in turn, answers the question of *what* the processes are but not necessarily *how* to perform them. That is, only the constituents expected to be part of the application process will be found on the Chinese Web site. In contrast, the USDS Web site tells both *what* the processes are and *how* to perform them. It specifies not only the expected constituents/parts but also the time when the constituents will be used, where to use them, and how to use them.

International Web users may find it frustrating to try to follow the visa application procedure as defined in the Chinese Web site, because the "procedure" does not specify a procedure at all. Instead, it names the steps and the "constituents" of the steps (the documents that are required). Again, this seems to follow the Confucian concept that if one knows the parts (i.e., their names), then one will know how to make use of those parts in the correct (i.e., harmonious) way. On the Chinese Web site, defining the constituents in the visa application procedure seems to serve two rhetorical purposes:

- To ensure the continuity of the entire procedure
- To filter out unnecessary information

As the discussion of Confucianism indicates, to restore and maintain social stability, individuals must follow the inevitable natural and social cycles. According to Confucianism, using correct names is conducive to the completing of a cycle. As noted earlier, one strategy championed by the *Shi Jing* is naming the constituents of a cycle. According to Confucianist principles, Chinese Web sites that provide users with the proper names of constituent parts associated with a process assume that this naming also provides users with a mechanism for correctly performing the related activities. Such an approach embodies the high-context notion of expecting individuals to intuit needed information from the context.

Confucianism and Chinese Web Design

My examination indicates that Chinese Web designers are still using Confucian strategy when creating professional communication documents, especially the documents that help an audience perform a task. According to Confucianism, a task is often a repeatable cycle: One begins at the beginning, then follows the steps, and finally stops at the ending. Individuals must follow the cycle as it exists to ensure success. Several factors influence the continuity of a task, including the various constituents/parts of a task and the person who performs the task. Performers may change, but the constituents—or the parts of an overall process—generally remain stable. Thus, continuity depends largely upon the constituents, and the constituents must be named correctly to maintain their correct use and the continuity of the overall cycle. The Chinese Web site seems to tie into this perspective of continuity and the maintenance of harmony via naming.

Naming the constituents also serves to filter out unnecessary information. One of the problems Web designers are facing today is that they may too easily display a huge amount of information on a Web site. Researchers have already pointed out that the information found on Web sites can be so overwhelming to users that they may not be able to locate relevant or needed information within an overall site (Barnum, Henderson, Hood, & Jordan, 2004; Spyridakis, 2000; Williams, 2000). To help Web users access or retrieve information, researchers have suggested that Web designers should select specific kinds of information to display on a Web site (Farkas & Farkas, 2002; Gregory, 2004; Spyridakis, 2000). One way to select such information is to choose the items that seldom change—information Web users are looking for, regardless of their particular circumstances.

In naming the constituents of the visa application procedure, Chinese Web designers seem to engage in selection of content information. The Confucian concept of naming things seems to provide one strategy for the Chinese Web designers, who display information such as visa categories, the application form, and the required documents—specific information useful to all Chinese visa applicants. In this way, Chinese Web designers represent a continuation of a much older cultural-rhetorical tradition.

Traditionally, Chinese professional communicators, including Web designers, stress content information, a point Barnum and Li (2006) raise while discussing the differences between Chinese and U.S. Web designs. The content information selected by the designers of the CFM's Web site emphasizes "commonalities" and denies any "particularities" of the present moment. That is, the content information—visa categories and forms—are equally useful to any Web user, no matter when, where, how, and why the Web user applies for a visa. (This is a classic case of naming representing stability.) Maybe the applicant is from Michigan, in the United States, and she wants to conduct a 7-day seminar at a Shanghai university. Another applicant is perhaps from a German electronic company, and he has been invited by a Chinese partnership to the opening ceremony of a new company at Beijing. From a Confucian rhetorical perspective, both applicants will find the visa information equally useful and relevant because both need the same constituents—visa categories and forms named on the Web—to complete the same process correctly. From a Confucian standpoint, to provide both individuals with the same names/categories of visa information is to provide them—and all other users—with the elements needed to complete that process correctly.

Such an approach, however, may frustrate non-Chinese Web users who are accustomed to a different rhetorical approach—especially Chinese visa applicants from low-context cultures such as the United States. As both Stewart and Bennett (1991) and Hall (1997) have pointed out, in low-context cultures, communication depends on connection between ideas and language. So, in low-context cultures, communicators always formulate ideas in very clear language. When providing a set of instructions, for instance, low-context culture communicators always provide task-oriented, step-by-step, actions for users to perform in order to finish a task. Thus, a Web user from a low-context culture may become frustrated when looking for an action-oriented procedure by which to apply for a Chinese visa, only to find a list of names and categories.

The strategy of naming the constituents is not limited to online documents. As Ding (2003) points out, Chinese print documents, particularly instructional manuals, also use this rhetorical strategy. For example, in guiding users to perform a task (e.g., assembling a water heater), such documents often name the various parts of the unit and reproduce the specifications of the unit instead of describing the actions that users need perform in order to install or maintain the unit.

Thus, both online and print documents represent a similar Confucian philosophy: Once users learn the correct names of the constituents, and once they have the correctly named constituents before them, they will easily figure out how to perform their task (see also Ding, 2003; Wang, 2000). Both online and print documents acknowledge "commonalities" and deny "particularities." The Chinese Web site, in selecting content information for display, prefers information "common" to all users (i.e., the naming of parts) and leaves out information useful only to certain individuals (i.e., using parts for specific vs. general purposes).

CONCLUDING REMARKS

In this chapter, I do not mean to claim that the Chinese Web designers of today simply pick up Confucian philosophy and apply it to their communication practice or that they set out to prove the efficacy of Confucian philosophical theory via their design. Rather, the modern practice of Web designing in China appears to dovetail with Confucian philosophy, and Confucian philosophy does help us uncover a pattern of meaning in the design of Chinese Web sites. Based on this, I present the following suggestions for international Web designers who develop online information for Chinese audiences and also suggestions for future research in this area:

- First, I would like to reiterate Rogers's (2008) suggestion that Web designers should put as much information as possible on a Web site, especially names of things and categories. These names and categories would help familiarize Chinese Web users with the content information they are looking for.
- Second, international Web designers should define all the important constituents in a procedure when they are providing instructions. Defining the constituents seems to guarantee a complete cycle through the procedure. Of course, more research needs to be conducted in order for us to learn more about Chinese culture and Chinese online culture.

Future research should address the following, related questions:

- Is the strategy of naming things a dominant feature on Chinese Web sites? Clearly, more Chinese Web sites need to be studied.
- Is the display of large amounts of information on a single Chinese Web page, as Rogers (2008) has pointed out, directly related to this strategy? Comparisons and contrasts need to be made between, say, the types of information displayed and the practice of naming things on a Web site.
- How usable is a Web page if it mainly names things and categories in a procedure? Usability studies need to be conducted on Chinese Web pages.

Although we need to learn more about the influence of Chinese culture on Chinese Web design, I hope that my Chapter helps international Web designers communicate more effectively with Chinese audiences and also make more informed decisions while developing online information for Chinese users.

APPENDIX 1: CFM WEB SITE (Taken on September 26, 2006 from http://www.fmprc.gov.cn/eng/ljzg/3647/3648/t18417.htm)

A Brief Introduction to Chinese Visa and A Brief Introduction to Visa Application Procedure

Chinese visa is a permit issued to a foreigner by the Chinese visa authorities for entry into, exit from or transit through the Chinese territory. The Chinese visa authorities may issue a diplomatic, courtesy, service or ordinary visa to a foreigner according to his identity, purpose of visit to China and passport type. Hereunder is an introduction to the ordinary visa and its application procedure.

The ordinary visas consist of eight sub-categories, which are marked with Chinese phonetic letters (D, Z, X, F, L, G, C, J-1 and J-2 respectively):

Visa D: issued to aliens who are to reside permanently in China. A permanent residence confirmation form shall be required for the application of Visa D. The applicant shall apply to obtain this form himself or through his designated relatives in China from the exit-and-entry department of the public security bureau in the city or county where he applies to reside.

Visa Z: Issued to aliens who are to take up posts or employment in China, and to their accompanying family members. To apply for a Visa Z, an Employment License of the People's Republic of China for Foreigners (which could be obtained by the employer in China from the provincial or municipal labor authorities) and a visa notification letter/telegram issued by an authorized organization or company are required.

Visa X: Issued to aliens who come to China for study, advanced studies or job-training for a period of six months or more. To apply for a Visa X, certificates from the receiving unit and the competent authority concerned are required, i.e., Application Form for Overseas Students to China (JW201 Form or JW202 Form), Admission Notice and Physical Examination Record for Foreigners.

Visa F: Issued to an applicant who is invited to China on a visit, on a study or lecture, business tour, for scientific-technological and cultural exchanges, for short-term refresher course or for job-training, for a period of no more than six months. To apply for a Visa F, the invitation letter from the inviting unit or the visa notification letter/telegram from the authorized unit is required.

Visa L: Issued to aliens who come to China for sightseeing, visiting relatives or other private purposes. For a tourist applicant, in principle he shall evidence his financial capability of covering the travelling expenses in China, and when necessary, provide the air, train or ship tickets to the heading country/region after leaving China. For the applicants who come to China to visit relatives, some are required to provide invitation letters from their relatives in China.

Visa G: Issued to aliens who transit through China. The applicants are required to show valid visas and on-going tickets to the heading countries/regions [that is, the countries they are heading to].

Visa C: Issued to train attendants, air crewmembers and seamen operating international services, and to their accompanying family members. To apply for a visa C, relevant documents are required to be provided in accordance with bilateral agreements or regulations of the Chinese side.

Visa J-1: Issued to foreign resident correspondents in China.

Visa J-2: Issued to foreign correspondents who make short trip to China on reporting tasks. The applicants for J-1 and J-2 visas are required to provide a certificate issued by the competent Chinese authorities.

In addition to providing the above-mentioned documents, an applicant is also required to answer relevant questions and go through the following formalities (with the exception of those stipulated otherwise by agreements):

Providing valid passport or a travel document in lieu of the passport.
Filling out a visa application form, and providing a recent 2-inch, bareheaded and full-faced passport photo.
Paying the visa fee

The overseas Chinese visa authorities are Chinese embassies, consulates, visa offices, the consular department of the Office of the Commissioner of the Ministry of Foreign Affairs in HKSAR, and other agencies abroad authorized by the Ministry of Foreign Affairs of China. If a foreigner intends to enter into, exit from or transit through the Chinese territory, he shall apply to the above-mentioned Chinese visa authorities for a Chinese visa. For further information, please consult the nearest Chinese visa authorities.

APPENDIX 2: USDS WEB SITE (Taken on October 5, 2006 from http://beijing.usembassy.gov/howtoapply.html)

How to Apply for a Visa

1) **Steps in Making an Appointment**
   a. **Buy a pre-paid PIN card from China CITIC Bank or visit the Visa Information Call Center website to purchase a PIN number online. PIN cards cost 54 RMB for 12 minutes of phone time or 36 RMB for**

8 minutes. Any unused minutes can be used at a later time or transferred to another party.

b. You may also pay the visa application fee (RMB 800) at China CITIC Bank. Please note that this fee is non-refundable and the original receipts are valid for one year.

c. Call the Visa Information Call Center to make an interview appointment or ask specific questions about the visa process. Please ensure that you know the following information: applicant's full name, passport number, ID number, contact information, purpose of travel, place of residence, whether s/he was refused before, etc. *All long distance charges are the sole responsibility of the caller, and all calls are generally answered in less than 30 seconds.*

Call Center Numbers:
Within China: 4008-872-333 (toll free)
(021) 3881-4611 (Shanghai local number)
Outside of China: (86-21) 3881-4611
Hours: Monday-Friday from 7:00 AM –7:00 PM
Saturday from 8:00 AM–5:00 PM
Pre-recorded visa information is available 24 hours a day, 7 days a week.

2) Complete Visa Application Forms

a. The English DS-156 non-immigrant visa application can be filled out online. Please print out all pages and submit them at the time of the visa interview.

b. Other non-immigrant visa application forms can be downloaded from the Embassy's website or can be obtained for free at a local branch of China CITIC Bank.

c. Glue one 2 × 2 inches color photograph to the *English* DS-156 form. The photo should be less than 6 months old, have a white background, and be a full frontal view of your face. Click for more information on Photo Requirements.

3) On the Day of the Interview

a. Bring all supporting documents with you for visa interview. For information on the required and suggested documents, please check the non-immigrant visa types.

b. Go to the Embassy or Consulate and Line up outside.

c. Go through the security check—Do NOT bring electronic devices, including cellular phones. Backpacks, suitcases, attaché cases, and strollers are also not permitted. Applicants should only bring documents relevant to the visa application.

d. Submit visa application forms and required documents at designated window, and then wait for fingerprint scanning and visa interview. Be prepared to wait 2 to 3 hours.
e. If your visa application was approved after the interview, your visaed passport will be mailed to you within 5 working days. Expedited pick up is available two working days after the interview date from the China Post Office located in the Chaoyang Government Visa Service Center near the entrance to the Visa Section of the Embassy. Special circumstances, such as the need for special clearances, incomplete applications, and/or fraud investigations may influence the processing of individual cases and delay visa issuance.

REFERENCES

Barnum, C., Henderson, E., Hood, A., & Jordan, R. (2004). Index versus full-text search: A usability study of user preference and performance. *Technical Communication, 51*(2), 185–206.

Barnum, C., & Li, H. (2006). Chinese and American technical communication: A cross-cultural comparison of differences. *Technical Communication, 53*(2), 143–165.

BOOC. (2008). Payment instructions for Beijing Olympic tickets phase 3 sales. Retrieved May 23, 2008, from www.tickets.beijing2008.cn/h/phase3_paymentinstruction.html

Chinese Embassy. (2006). General procedures for application. Retrieved October 3, 2006, from www.china-embassy.org/eng/hzqz/gpro/t230371.htm

CFM. (2006). Travel to China. Retrieved September 26, 2006, from http://www.fmprc.gov.cn/eng/ljzg/

Confucius. (1979). *The analects*. (D. C. Lau, Trans.). New York: Penguin Books.

Ding, D. (2003). The emergence of technical communication in China—*Yi Jing (I Ching)*: The budding of a tradition. *Journal of Business and Technical Communication, 17*(3), 319–345.

Farkas, D., & Farkas, J. (2002). *Principles of Web design*. New York: Longman.

Fung, Y. (1997). *A short history of Chinese philosophy* (3rd ed.). (D. Bodde, Trans.). New York: Free Press.

Fung, Y. (2003). *A history of Chinese philosophy: The period of philosophers.* (D. B. Bodde, Trans.). Princeton, NJ: Princeton University Press.

Gao, S., & Handley-Schachler, M. (2003). The influences of Confucianism, Fengshui and Buddhism in Chinese accounting history. *Accounting, Business and Financial History, 13,* 41–68.

Gregory, J. (2004). Writing for the Web versus writing for print: Are they really so different? *Technical Communication, 51*(2), 276–285.

Hachigian, N. (2001). China's cyber strategy. *Foreign Affairs, 80,* 118 133.

Hagemann, J. (1986, May). Confucius says: Naming as social code in ancient China. In *Annual Meeting of the Conference on College Composition and Communication.* New Orleans, LA. (ERIC Document Reproduction Service No. ED272892)

Hall, E. (1997). Context and meaning. In L. Samovar & R. Porter (Eds.), *Intercultural communication* (pp. 45–54). Belmont, CA: Wadsworth.

Herr, R. (2003). Is Confucianism compatible with care ethics? A critique. *Philosophy East and West, 53*(4), 471–489.

Legge, J. (Ed. and Trans.). (1993). *The I Ching: The Book of Changes.* New York: Dover Publications.

Li, W. (2007, January 24). Internet users to log in at world. *China Daily*, p. 2.

McCool, M. (2006). Information architecture: Intercultural human factors. *Technical Communication, 53*(2), 167–183.

MOST. (2007). Call for papers. Retrieved March 17, 2007, from www.most.gov.cn/en/anouncementboard/200606/t20060616.34260.htm

NTAC. (2006). China travel. Retrieved October 05, 2006, from http://www.old.cnta.gov.cn/lyen/

Owen, S. (1996). Foreword. In A. Waley (Trans.) & J. Allen (Ed. & Trans.), *The Book of Songs: The ancient Chinese classic of poetry* (pp. xii–xxv). New York: Groves Press.

Owen, S. (2000). Reproduction in the *Shijing* (Classic of Poetry). *Harvard Journal of Asiatic Studies, 61*(2), 87–315.

Rogers, K. (2008, May). The culture of China's Internet. *Intercom*, 10–13.

State Council. (2006). *Maintain a harmonious society to expedite positive changes in China's economy*. Retrieved October 4, 2006, from http://www.gov.cn/jrzg

Spyridakis, J. (2000). Guidelines for authoring comprehensible Web pages and evaluating their success. *Technical Communication, 47*(3), 359–382.

St.Amant, K. (2005). A prototype theory approach to international Web site analysis and design. *Technical Communication Quarterly, 14*(2), 73–91.

St.Amant, K. (2007). Online education in an age of globalization: Foundational perspectives and practices for technical communication instructors and trainers. *Technical Communication Quarterly, 16*(1), 13–30.

Stewart, E., & Bennett, M. (1991). *American cultural patterns: A cross cultural perspective*. Yarmouth, ME: Intercultural Press.

U.S. Embassy, Beijing. (2006). How to apply for a visa. Retrieved October 5, 2006, from http://beijing.usembassy.gov/howtoapply.html

USDS (2006). Travel. Retrieved October 5, 2006, from http://www.state.gov/travel

Wang, Q. (2000). A cross-cultural comparison of the use of graphics in scientific and technical communication. *Technical Communication, 47*(4), 553–560.

Williams, T. (2000). Guidelines for designing and evaluating the display of information on the Web. *Technical Communication, 47*(3), 383–396.

CHAPTER 6

What We Have Here Is a Failure to Communicate: How Cultural Factors Affect Online Communication Between East and West

Carol M. Barnum

CHAPTER OVERVIEW

This chapter presents an analysis of rhetorical differences in communication between Asian and North American writers. Using the work of Hall and Hofstede as the basis for analysis, the chapter presents several examples of business correspondence (letters sent by e-mail) from Asian (mostly Chinese) senders to me in the United States. An additional example of business communication, which is more broadly intended to reach an international English-speaking audience, is posted in English on a Chinese Web site as a letter of introduction to potential business partners. Ways in which the analyses of these letters can be useful to classroom teachers and trainers are suggested, as well as ways in which these teachers and trainers can gain more understanding of the rhetorical cultural bases of computer-mediated communication between Asian and U.S. correspondents.

It has become a cliché to say that the world is a global village, largely because of the Internet, but this situation provides the means for easy access to almost anyone almost anywhere in the world at almost any time. Although such ease of access has changed the speed and volume of communications being sent, it has not necessarily improved the effectiveness of communication. With e-mail increasingly becoming the medium of choice for communication worldwide, cultural-rhetorical conventions can and increasingly do block the ability of both sender and receiver to understand each other. Because the cultural differences can be so extreme when communicating between East and West, these communication efforts can be the most problematic.

Through an analysis of several e-mail communications from Asian (mostly Chinese) senders to me (a U.S. recipient), I discuss the cultural basis for the communication differences based on Edward T. Hall's descriptions of high-context and low-context cultures (1989) and Geert Hofstede's five cultural dimensions (see Hofstede & Hofstede, 2005). I extend the discussion to include a letter of introduction (in English) on a Chinese Web site. I conclude with recommendations for teachers of business and technical communication courses on ways to learn more about the cultural bases for the communication differences between Asia and North America, so as to prepare our students to effectively interpret messages from these cultures and create culturally appropriate messages.

LAYING THE GROUNDWORK FOR ANALYSIS

According to recent data available about e-mail usage, almost 97 billion e-mails were estimated to have been sent daily worldwide in 2007, up from 36 billion person-to-person e-mails sent daily in 2005 (ITFacts, 2007). With the increasingly easy access to e-mail addresses via Web sites hosted by businesses and universities comes an associated increase in communication traffic from strangers. Because of the cultural differences evident in e-mail correspondence between Asia and North America, a message crafted by a sender in one cultural context may not be understood correctly by the receiver in another cultural context; and the resulting response, when there is one, may disappoint the sender. In comparison with the standard business letter, in which certain issues of "noise," such as spelling or typing errors, word choice, and tone, may diminish the effectiveness of the message, the cultural issues reflected in international e-mail correspondence are likely to add to the noise and thereby decrease the effectiveness of communication, resulting in "a failure to communicate." Many of the popular business and technical communication texts are addressing cultural issues that affect communication, including e-mail messages in business (Bovée & Thill, 2005; Guffy, 2003; Houp, Pearsall, Tebeaux, & Dragga, 2002). However, these topics, typically treated in a chapter or section in a text, may not provide sufficient guidance on how to improve the likelihood of a satisfactory cycle of communication between international/intercultural senders and receivers of messages.

My own experience as a business and technical communication instructor and as a speaker at international conferences and in training activities, most frequently in Asia, has caused me to be the recipient of e-mail requests for help, support, and advice from both little-known acquaintances and complete strangers. My interest in international technical communication led me to develop a course in our graduate program, and has deepened my appreciation for the ways in which the studies conducted by Hall (1989) and Hofstede (see Hofstede & Hofstede, 2005) and the work of contrastive rhetoricians such as Kaplan (1966) and Connor (1996) provide a useful basis for understanding the rhetorical and structural differences so clearly present in much e-mail and other business communication between Asian and North American correspondents. Using personal examples allows me to establish the context for the communication. My approach is to analyze the rhetorical style of several Asian letters I have received, plus an open letter of introduction on a Chinese Web site, interpreted through the lens of the work of Hall and Hofstede. Where relevant literature supports my analysis of the content from a cultural perspective, I have provided such support. However, in some cases, my analysis is speculative; for no matter how much I study Asian culture, I will remain an outsider, with my cultural blinders very likely restricting my perspective. The purpose of the analysis of the samples in this chapter is to add to the limited but growing conversation on this topic.

APPLYING THE WORK OF HALL AND HOFSTEDE

Although there are a number of cultural and social anthropologists doing work in culture studies, Edward T. Hall and Geert Hofstede are arguably the most commonly cited. Hall (1989) categorizes cultures as existing along a continuum from high-context to low-context, with Asian countries such as Japan and Korea being at the high-context end, and Western countries such as Switzerland and Germany being at the low-context end. China ranks highest on the high-context end; and North America, although not the lowest, ranks on the lower-context side of the continuum, placed 11th out of 14 cultures that Hall lists by way of example.

For those not familiar with the terminology, Hall characterizes high-context cultures as those in which meaning and context are interconnected, in such a way that communication depends as much or more on the unspoken, the subtext of a message, than on the words used. In contrast, low-context cultures derive meaning from the words themselves, expecting messages to provide clarity and explicitness. Low-context cultures prefer direct message structure and straightforward diction. This contextual framework reflects the approach used in teaching effective communication strategies in U.S. academic and business settings. High-context cultures view such directness as abrupt, demanding, and intrusive, preferring an indirect approach, which is polite, often deferential, and formal in tone. Such an approach allows the reader to infer meaning from the message. Such a message structure also pays homage to the importance of "face," in providing a means to craft messages that protect the receiver, as well as the sender, from a loss of face through any number of potential pitfalls, including making a mistake or pointing out an error, making a demand, showing a lack of respect for someone in a senior position, even making a recommendation that might suggest a lack of respect for the receiver's decision-making capabilities and responsibilities.

Although Western writers must also consider face-saving strategies in their communication, the Chinese concept of face (*mianzi*) is far more complex and is a critical component of communication. As Cardon and Scott (2003) report, the Chinese language has hundreds of phrases involving "face" and "face-related behavior." Among its complex characteristics are the following understandings:

- Face can be measured—higher status equates to having more face.
- The amount of face can be altered—face can change over time; face can be lost or restored and can be increased through business relationships.
- Face can be exchanged—face is a form of "currency," which means that a person with a lot of face may be asked to serve as a go-between for another person with less face; this type of arrangement is frequently used to set up introductions between a person with less face and another with more face.
- Face is mutual—if one person loses face, all in the group lose face; all group members share the responsibility to enhance the status of the group; commonly exhibited by members of a group describing the accomplishments of other, higher-status members.
- Face is influenced by others—one "gives face" to others in a business relationship as part of the process of building trust and establishing credibility; this is exhibited by publicly praising the other in recognition of status, position, or managerial effectiveness.
- Losing face has serious implications—face-losing actions include displaying anger, directly refusing requests, failing to honor requests, acting aggressively, and failing to show appropriate respect for the status of others (Cardon & Scott, 2003, pp. 12–14).

Hofstede also addresses issues of face as critical to two of his five dimensions of culture. Building on the work of Hall, Hofstede contextualizes high-context cultures as *affiliation* cultures and low-context cultures as *achievement* cultures. Hofstede's discussion of the five dimensions of cultures can be summarized briefly as follows:

Power Distance	The extent to which the less powerful members of institutions and groups expect and accept that power is distributed unequally.

Individualism vs. Collectivism	Individualism focuses on the individual, with the expectation that everyone looks after him- or herself, whereas collectivism exhibits the expectation that from birth onward people are integrated into strong, cohesive groups. Includes concern for face.
Uncertainty Avoidance	The extent to which members of a culture feel threatened by uncertain or unknown situations.
Masculinity vs. Femininity	Refers to gender roles, not physical characteristics, with masculine cultures exhibiting assertiveness and a focus on material success, and feminine cultures exhibiting modest, tender, caring qualities with a focus on relationships.
Long-term vs. Short-term Orientation (also called the Confucian work dynamism)	Reflects Confucian principles of respect for tradition and social and status obligations, virtue in future rewards; includes concern for face.

The following table shows the relative differences between China and the United States for each of the five dimensions. The numbers reflect Hofstede and Hofstede's (2005) index, which ranges roughly from 0 to 100, with the higher numbers being more closely correlated to the dimension listed in the table. China scores high on power distance and has the highest score on long-term orientation, scores relatively high on uncertainty avoidance, close to the United States on masculinity, and extremely low on individualism. The United States' scores place it generally on the opposite side of the scale from China, with the most extreme differences reflected in individualism (U.S. has the highest score of all the 76 countries represented in the index) and long-term orientation (China has the highest score).

Cultural Dimension[a]	United States	China
Power Distance (104)	40	80
Individualism (91)	91	20
Uncertainty Avoidance (104)	46	30
Masculinity (95)	62	66
Long-term Orientation (118)	29	118

[a]The highest score attained by any country for each dimension is indicated in parentheses.

IMPACT OF CULTURAL DIFFERENCES ON COMMUNICATION STYLE

When we compare China and the United States on the basis of Hofstede's cultural dimensions and the corresponding contextual reference point (high-context or low-context), the impact of each culture's differences on communication can be profound. Kaplan was the first to document this cultural basis for differences in communication style in his 1966 article on contrastive rhetoric, in which he discovered that the writing patterns of Chinese and other international students were profoundly different from those of American students, and that these patterns were deeply engrained in culture. More recent interpretations of Kaplan's work can be found in the work of Connor (2002). In *The Geography of Thought: How Asians and Westerners Think Differently . . . And Why* (2003), Nisbett characterizes the differences as being largely based on the Western Aristotelian/ Cartesian worldview versus the Eastern Confucian/Taoist worldview. For example, the Chinese see the natural world as "a mass of substances rather than a collection of discrete objects. Looking at a piece of wood, the Chinese philosopher saw a seamless whole composed of a single substance. . . . The Greek philosopher would have seen an object composed of particles" (p. 18).

Another contrast described by Nisbett (2003) is in attitudes toward argument: "Greeks were independent and engaged in verbal contention and debate in an effort to discover what people took to be the truth. They thought of themselves as individuals with distinctive properties, as units separate from others within the society, and in control of their own destinies" (p. 19). Their speaking and writing reflected this philosophy, with praise given to inventive argument and clearly stated positions on issues. Chinese social interaction, in contrast, was "interdependent and it was not liberty but harmony that was the watchword—the harmony of humans and nature for the Taoists and the harmony of humans with other humans for the Confucians" (p. 19). The Chinese philosopher would see a family of interrelated members; the Greek philosopher would see a loose collection of individuals with unique characteristics. Nisbett sums up the differences in this way: "I once asked a Chinese philosopher why he thought the East and the West had developed such different habits of thought. 'Because you had Aristotle and we had Confucius,' he replied" (p. 29).

Such disparate philosophical views of the world have led, naturally, to disparate approaches to communication. Western, in this case United States, writing strongly favors deductive (direct) organization with a clear purpose statement, direct requests, concrete conclusion, and specific recommendation. Eastern, in this case Chinese, writing favors exactly the opposite: inductive (indirect) organization with the purpose for writing subordinated to or assumed from the context; it is characterized by indirect requests, a circular statement of ideas, and an economy of language called "synthetic" writing in its synthesis of ideas into a few key words (Barnum & Li, 2006; Zhu & St.Amant, 2007). The reader is expected to know how to extract the meaning. In the rare instances in which recommendations are stated, these are shaped in the form of "suggestions" so as to avoid an appearance or tone of presumption or arrogance. Chinese communication uses an indirect style that emphasizes relationships (with ample use of the pronoun "we"), recognition of status differences, and language rich in metaphor, which U.S. readers might unflatteringly characterize as verbose, inefficient, dishonest, and vague. U.S. communication uses a direct style that emphasizes personal responsibility (with ample use of the pronoun "I"), individual effort in attaining success, and expectation of personal reward, which Chinese readers might unflatteringly characterize as abrupt, demanding, and boastful.

The inductive or indirect, sometimes circular writing pattern of the Chinese is recognized as an issue for those teaching technical writing in China. Li Huilin, a colleague who teaches English and technical writing in Kunming, China, reports that he could provide "hundreds of examples of the inductive writing pattern favored by Chinese students" (Barnum & Li, 2006, p. 153). And he is not alone. Research conducted by Professor Yang Yuchen, vice dean of the School of Foreign Languages at Northeast Normal University, confirmed that

> the lack of a topic sentence and the use of inductive order are cited as two significant problems in teaching the Western style of writing to Chinese students. In an Internet-based technical English learning program offered by Shanghai Communication University, the American professor grading the Chinese students' papers documented that 30% of the compositions have problems with the top portion and that the topic sentence is the critical problem for the Chinese learner to pay attention to in writing English (personal interview, August 4, 2005, Changchun, China, cited in Barnum & Li, 2006, p. 153).

Duan and Gu (2005) report the same problem with overseas Chinese students writing research papers at a U.S. university. They attribute the problem to "Chinese students' lack of familiarity with Western rhetorical patterns; instead [the students] wrote following traditional Chinese rhetoric." They conclude that Chinese students need an introduction to "the rhetorical features of Western technical communication" (p. 438).

ONLINE INTERCULTURAL COMMUNICATION ISSUES

As Ray Archee (2003) writes in his column on computer-mediated communication in *Intercom*:

> Most professionals would agree that the Internet enables us to communicate more effectively with our colleagues, both locally and internationally. . . . But we often overlook one area of computer-mediated communication (CMC): How do cultural differences affect successful online communication? Does Western culture's informal and direct use of CMC technologies conflict with the way other cultures use these technologies? Or has the world become a homogeneous community with each country indistinguishable in terms of its communication behavior online? (p. 40)

These questions are addressed in a number of studies reported in 2005 in a special section on "Culture and Computer-Mediated Communication" in the

Journal of Computer-Mediated Communication. Several of the studies focus on Web sites (advertising/marketing and universities) and arrive at the shared conclusion that greater trust is associated with Web sites whose graphical elements are localized to suit the expectations of the particular culture (Callahan, 2005; Hermeking, 2005). *The Culturally Customized Web Site* (Singh & Pereira, 2005) and *Beyond Borders: Web Globalization Strategies* (Yunker, 2003) support the need to localize a Web site to suit a culture and present numerous ways to do so for particular cultures. Another study (Faiola & Matei, 2005) demonstrates that a Web site designed by a Web designer from a particular culture for that culture (China for Chinese users and U.S. for U.S. users) improves the efficiency of users from the culture when they are performing information-seeking tasks. As the focus of these studies in the special section is on applying the work of Hall and Hofstede to determine its applicability to computer-mediated communication (CMC), the editors summarize the findings of these studies by saying that the articles "help to define more clearly those domains of online intercultural communication research that are well served by Hall's and Hofstede's frameworks, and those that are more fruitfully examined using alternative frameworks" (Ess & Sudweeks, 2005, para. 3).

Other studies examine issues of self-disclosure in CMC, comparing this medium to face-to-face (FTF) communication. One such study (Yum & Hara, 2005) compares issues of trust and disclosure in CMC versus FTF meetings. Although this study addresses the lack of research on how people use CMC as a "relational communication channel in different cultures, (para. 2)" the study limits its investigation to a 10-minute survey of CMC users' attitudes within their own respective cultures as they build relationships with others in their own culture. An earlier study (Ma, 1996) pairs North American and East Asian students for a CMC study of interpersonal relationships. The earlier study finds that the North American students perceived their East Asian partners as indirect and insufficiently self-disclosing, despite the fact that the East Asian students felt that the use of CMC provided a way for them to be much more self-disclosing and direct than they would be in face-to-face communication. In fact, the East Asian students characterized their approach as just as self-disclosing and direct as that of their North American communication partners. Yum and Hara suggest that the definition of directness is "subject to cultural variation" (2005), which this study seems to indicate.

Another study (Durham, 2003) looks at language preferences in CMC. This study reports that in a trilingual country like Switzerland, medical students on an e-mail mailing list chose English most often for publicly shared messages to all, but switched to their native tongue for individual communications, reflecting greater comfort in their native language, despite their fluency in another language. With each new study, we add to the small but growing body of literature about CMC. Still, little work has been done on the cultural influences reflected in e-mail when it is used as a business-related CMC.

E-MAIL CORRESPONDENCE FOR BUSINESS

The growing use of e-mail worldwide has changed the way business is typically conducted, with e-mail largely replacing the letter as the standard mode of business communication. The ease of drafting and sending e-mail has created a general state of e-mail "overload" so pervasive that articles and even a popular book have taken up the topic with strategies on how to handle the stress associated with the seemingly never-ending stream of e-mail that the typical business person must handle every day (see Fry, 2007; Sandberg, 2006; Shipley & Schwalbe, 2007).

An article in the *Journal of Business and Technical Communication* (Thomas et al., 2006) addresses the nature of e-mail overload, its root causes, and the characteristics that contribute to the resulting sense of stress. What is interesting to note, in light of the issues with regard to using e-mail in international communication, is that the unique characteristics of e-mail that make it the current medium of choice are the same characteristics that result in communication problems when used in intercultural exchanges. These characteristics of the medium include (Thomas et al., pp. 254–255) the following:

- Asynchronous communication—it does not require the sender and receiver to be present for communication to occur.
- Written messages—aside from providing a record of the communication, the specific nature of the communication is that typically it takes on the characteristics of informal, spoken conversation. Also the common occurrence of short, informal strings of messages resembles the conversational turn-taking that occurs in face-to-face meetings.
- Multiple recipients in a single action—senders can distribute the same information to a large number of recipients at once.
- Parallel structure memory—messages to be stored, retrieved, and forwarded, which allows for social memory and establishes accountability.

Why do these characteristics of e-mail exacerbate the problems for cross-cultural communication? Let us examine each one in turn.

Asynchronous Transmission

The ability to communicate at the convenience of one's own schedule and time zone is clearly a benefit of e-mail on one level, in light of the significant time zone differences between Asia and the United States. However, Asian cultures have a different perspective on time than North American cultures, another topic of research explored by Hall (1989, pp. 17–19). Asian cultures are characterized as both high-context and *polychronic* (P-time) cultures; and North American cultures are characterized as low-context and *monochronic* (M-time) cultures. P-time cultures view time as simultaneous and concurrent, with many things happening at once. Priorities have a tendency to shift, as a commitment to people's needs takes greater precedence than adherence to a predetermined schedule. In contrast, M-time cultures view time as sequential and linear, with time commitments and scheduling seen as critical.

For P-time cultures, e-mail seems to support the need for connecting and building relationships. The authors of a study of Internet users in Singapore (Lee, Tan, & Hameed, 2005) find that the intuitive notion of a positive correlation between P-time users and the Internet is supported, although they find no corresponding correlation in terms of the "displacement hypothesis," which posits that increased use of the Internet displaces something else, such as television viewing, radio listening, or newspaper reading. Although Internet usage in Singapore increased from 5.2 hours in 1999 to 10.5 hours in 2005, the time spent on the other activities was unchanged. Lee, Tan, and Hameed conclude that there is a lack of change in usage because

> [P]olychrons are adept at multi-tasking, and in restructuring and reorganizing activities. They are flexible and comfortable with uncertainty. Their lives often seem more chaotic due to their habit of juggling several tasks at once and frequently changing schedules. Monochrons, on the other hand, prefer to focus on one task at a time. They hate disruptions and changes to plans. They appear focused and efficient. (para. 34)

Despite the fact that, as the authors point out, there are several limitations in terms of their findings, in that the study focused on home use and did not include cross-cultural comparison data, their findings may support the idea that P-time users adapt more easily to one more communication medium affecting their time; whereas M-time users feel increased stress due to the increasing use of e-mail and their resulting lack of control over their time.

However, when it comes to certain business transactions, P-time cultures find e-mail is less useful as a business tool when face-to-face meetings are possible. Regardless of the time and cost involved in traveling to such face-to-face meetings, such meetings are more highly valued than e-mail for the cultural benefits they provide: fostering and maintaining relationships, demonstrating the importance of the event being discussed or celebrated, and supplying the all-important context for the interpretation of contextual cues. A survey of the communication preferences of virtual team members in Korea (Lee, 2000) showed a strong preference for e-mail with peers, whereas the preference shifted to a different medium—one showing more respect—when communicating with a manager. When a face-to-face meeting was possible, it was viewed as optimal, with fax or overnight letter as second choice.

Written Messages

In low-context, M-time cultures, written messages are used to spell out issues, cement understanding, and serve as concrete evidence of actions, agreements, and so forth. The efficiency of e-mail supports the aims of M-time cultures. In high-context, P-time cultures, written documentation, such as contracts or agreements contained in or attached to an e-mail, may not be viewed as set in stone. To complicate matters for the Asian e-mail correspondent, who is typically writing in a nonnative language and within a formal communication context, the seemingly casual writing style of the American e-mail correspondent may suggest an inappropriate tone to the Asian correspondent. This concern for tone and the need to show proper respect is reflected in the survey and interview data collected on the Korean virtual team members' preference for media other than e-mail when communicating with their managers. As the study's author points out, "Confucian cultural protocol influences Koreans when they are trying to use e-mail to communicate with their seniors because it forces them to illuminate the code of respect usually embedded in their subconscious" (Lee, 2000, p. 197).

Multiple Recipients, Built-in Memory, and Accountability

The convenience of being able to send a message to multiple recipients at the same time certainly aids communication among members of globally dispersed teams and, as such, is prized by the North American correspondent. However, the Asian sender or recipient may be unsure as to how to write or respond to multiple recipients when issues of status and hierarchy raise the problem of how to address readers at different status levels.

E-mail's features of built-in memory and the resulting accountability may cause problems from an Asian perspective for two reasons: (1) what was agreed to earlier does not necessarily hold later; and (2) accountability to the group is more highly prized than accountability to the individual. Because Asian affiliation cultures more commonly place a higher value on the relationship than on whatever language may be in an agreement or contract, it is generally understood that the strength of the relationship provides for accommodation as the need arises, and such accommodation is a reaffirmation of the relationship.

ANALYSIS OF E-MAIL CORRESPONDENCE

I now present three e-mail letters that I received from Asian correspondents. These letters are analyzed on the basis of certain cultural factors present in the organization, tone, and approach used in them. The fourth example is also a letter, written in English and available on a Chinese Web site. In this fourth example, the correspondent is writing to any English-speaking readers with an interest in developing a business partnership. In discussing these examples, I posit that the differences from U.S. patterns reflected in the communication patterns used by the Asian writers of business-related messages indicate a deeply ingrained cultural pattern that can be understood to some degree by applying the work of Hall and Hofstede.

The use of these examples is not meant to suggest that they are representative of a particular cultural style but to show how these particular letters reflect cultural values and can be understood from the cultural framework of the correspondent. The same approach could be used to interpret correspondence from an American/North American correspondent to an Asian recipient. An excellent analysis of the rhetorical patterns found in a Chinese letter (high-context culture) and the ways in which the letter poses problems of understanding when read in translation in a low-context culture like the United States or Northern Europe can be found in Campbell (1998). Campbell describes the issues that low-context cultures (those of Americans and Northern Europeans) have with the Chinese letter and contrasts this with reactions to the letter rewritten for low-context readers. However, in this chapter it not my intention to compare and contrast the cultural values reflected in correspondence from Asia to the United States and from the United States to Asia. The style used in American business communication is very likely to be well understood by the readers of this chapter. The style of the Chinese and other Asian examples that follow is less likely to be readily understood. Thus, the focus of my analysis is on the letters from Asia.

Example 1: E-Mail Letter from Chinese University

Sent by the director of the International Affairs Office of a Chinese university I had visited, the letter uses a familiar Chinese pattern, which I have seen in most correspondence I have received from Chinese writers. The structure is indirect (inductive). The tone is formal. The only personalized touch is the opening greeting: "How are you doing?" which may hint at a relationship. The salutation—"Dear my Colleagues"—indicates that the letter is being sent to a number of people. The indication of multiple recipients suggests that the writer sees all who receive the message as part of the same group.

The main contents of the letter address the formation of a new university out of the merger of several colleges. Factual information is presented and documented to instill credibility and official sanction, characteristics that further increase the formality and authority of the letter. The formation of the new university is placed in the context of supporting the future prosperity of China, as appropriate within the context of a Confucian/affiliation culture. The inductive structure places the purpose of the letter—the request for a letter of congratulations in honor of the occasion—in the second to last paragraph, first previewed in one sentence and then formally stated in the following sentence. The request does not make any direct reference to the relationship, but the meaning can be inferred.

The responsibility is on the receiver to respond as a demonstration that the relationship is valued. To do otherwise would result in a significant loss of face for both parties and would very likely jeopardize the relationship. Please note that the real names of institutions and individuals have been changed.

Subject: [Chinese] University Founded
Dear my Colleagues,

How are you doing?

I have the honor to inform you that [Chinese] University of Science and Technology, [Chinese] Medical College and [Chinese] Teacher's College were merged and formed [Chinese] University, on the date of August 18, 2001, according to the notice of Document No. 104 [2001] issue[d] by the People's Government of [Chinese] Province. It is no doubt that the formation of [Chinese] University will not only hasten reform steps and optimize structure of the layout of higher learning education in [Chinese] Province, and also improve education quality and school-running efficiency in [Chinese] Province, and serve provincial economy constructions in modernization better.

At the same time, Prof. Yang Yufei [named changed] is honorably appointed as the first term of the President of [Chinese] University.

A ceremony hanging out a shingle of the newly-founded [Chinese] University is going to be held on 28 October 2001. On behalf of the university, I have a request to make of you. Would you please kindly send us a congratulatory message on the [Chinese] University's formation and our new-appointed President?

I would appreciate you cordially for your kind support. I hope that we can receive the message from you before the holding of the ceremony.

With best regards,
[name deleted]
Director
International Office

My reply follows. In it, I attempt to adopt the same formal language and style as used in the letter from China, so as to convey my understanding of the significance of the event. I also recognize the importance of speaking for my university (which has higher authority and status than do I), despite the fact that I have been the only representative from my university to visit the Chinese university. Although I understand that it would be more fitting for someone in higher authority to speak for my university, that is not feasible, so I emphasize my title and degree, and fax and mail my reply on letter-head stationery. My reply takes into consideration the face-giving nature of the communication.

Congratulatory message from Professor Carol M. Barnum, Ph.D., Professor of Technical Communication at Southern Polytechnic State University, Marietta, GA, USA

In honor of the occasion of the inauguration of the newly-founded [Chinese] University and its newly appointed President Professor Yang Yufei [name changed], I offer my hearty congratulations on behalf of my university, Southern Polytechnic State University in Georgia.

As I have personally visited your campus on several occasions and have always been impressed with the hospitality and great evidence of interest in higher education, I can highly commend you on this latest, and most important step in consolidating resources and building strength as a university.

My best wishes for your continued success in your future educational endeavors.

Sincerely,
Carol M. Barnum, Ph.D.
Professor of English

Example 2: E-Mail Letter From Singapore Technical Writer

I received this e-mail letter from someone I had not met, but who had a relationship with someone else I allegedly met at a conference in India. The writer cites details of the conference to confirm, perhaps, that she has her facts in order, but she does not cite the name of the friend who attended, perhaps because she knows that the friend did not actually introduce herself/himself to me or perhaps because she has concluded that I will not recall having met the friend, if I did. Thus, she tries to establish her credibility by citing the details of the conference to demonstrate the legitimacy of the connection via the unnamed go-between.

From the writer's name (which has been changed here), I speculate that the writer is of Indian-Chinese descent; she indicates that she is writing from Singapore, which has strong Chinese and Indian communities. She writes to request information, but more than just information, she requests a recommendation from me as to which of two universities she should attend in the United States. The letter is the vehicle by which she presents herself. The structure is inductive with the purpose coming very late in the letter. After establishing her identity she takes pains to establish her credibility before making her request for advice by demonstrating what she knows about the career of technical communication and her work experience in preparing for this career. My interpretation of the significance of including these facts is that it is to persuade me to take her case seriously and to respond. Her English language proficiency suggests a certain level of comfort with colloquialisms, such as "leaps and bounds," and a sophisticated English vocabulary, reflected in words such as "dearth" and "irrespective."

Her first question—What are the essentials?—is rhetorical, in that she uses the question to set up her fact-based answer. Her later questions suggest to me that she has the expectation of a similarly fact-based response from me.

Subject: Graduate Studies

Dear Ms. Carol M. Barnum,

I, Usha Wu [name changed], 26-years old, with a Master's Degree in Political Science, would like to introduce myself as an ambitious lady with good communication skills, both written and verbal.

After four years, I took a break from the mainstream journalism in the year 200_ and joined IT industry for a career in Technical Writing.

Associated with information technology (IT), for the past one year and more as a Website Writer, I have come across various types of technical writing, internal documentation, product literature, user guides, online help, manuals, installation instructions and more. Basically, being a journalist by profession, there have been times when I felt that I can do a better professional job in the field. While the importance of technical writing is acknowledged universally, the specialization and skills one needs are not exactly defined. What are the essentials? Language, programming, planning and positioning of what the organization provides, crisp presentation or all put together.

With IT advancing all over the world by leaps and bounds, there is a dearth of technical writers who can bring in quality and content for a specific job. Writing has always been a pleasure for me and I see this current phase both as an opportunity and challenge for enhancing my abilities and advancement of career prospects.

Keeping these in view, I secured admission into the Department of Technical Communication, [X State University, City, State, and URL], and into the Master of Technical Writing Program at [Y University, City and URL].

I am unable to decide my choice over the Universities. A friend of mine who attended your presentation at the STC India chapter, 3rd annual conference, Bangalore, December 7–8, 2001, suggested that you could be the right person to help me choose the best out of the two universities.

Based on the academics (course curriculum) and academics alone irrespective of all other factors like fee, location, etc., could you please tell me, which school should I go?

Looking forward to your support.

Sincerely,
Usha Wu
Singapore

I have used this letter for discussion purposes in many of the cross-cultural communication workshops and seminars I have presented in such culturally diverse locations as China, New Zealand, Canada, and various places in the United States. The comments I have received from participants have broadened my own understanding of the cultural issues at work here. The main points are these:

- After addressing me by my full name, she introduces herself by name, presenting her age and the fact that she holds an advanced degree, which is perhaps meant to demonstrate her maturity and advanced level of education, supported by her statement that she is ambitious. The reference to being ambitious subtly suggests the purpose of the letter.
- The second paragraph suggests a connection through our shared interest in technical writing. The third paragraph establishes her credibility from her prior career and her knowledge of the IT field (which she meticulously abbreviates for future use as an initialism).
- In the third paragraph, she demonstrates her understanding; yet she suggests the need for clarification, in ending the statement with "or all put together."
- In the next paragraph, she presents the information about having been accepted by two U.S. universities, and includes the name of each degree program, the location of the university, and the URL for each (all deleted from the letter). Perhaps the inclusion of the URLs for the universities is a subtle way of suggesting that I should review the two programs before responding to her, or perhaps it suggests that she has already reviewed them and is knowledgeable at a certain level about the programs. If the former, she is not making a direct request that I review the programs at the universities, as this would be too assertive; instead, she provides me with the information so that I can choose to do so myself.
- In the next paragraph, she states her problem of not being able to make a decision followed by the reference to her connection with a friend who heard me speak at a conference. Using the persuasive tactic of telling me that the friend suggested I was the *right person* to help her make the best choice, she appeals to me to respond through this connection and through an implied sense of duty or obligation.
- In the next paragraph, she describes the criteria I should use in making the choice for her. She ends by reinforcing her desire to receive my "support."

As participants in my seminars always point out, this request puts an enormous burden on me as the recipient of this e-mail letter—from a stranger—to shape an important career move by helping her make the right choice. Although she does not know me, she clearly respects the advice of her friend in recommending me and she leans on that connection in hope of soliciting the right response. Western participants in my seminars also point out that they would very likely not read the lengthy e-mail all the way to the end, because they would quickly see that they do not know the sender and also because the organizational pattern is inductive, requiring careful reading of a great deal of the message before arriving at the request. Inevitably, they want to know how I replied.

Although I did not keep a copy of my reply, I recall the gist of it, which was this: I commended her for her desire to continue her education and complimented her on her understanding of the important role of technical communication in the IT industry and on being accepted at two U.S. universities. Although I was tempted to ask her whether she had given consideration to the graduate program at *my* university, I refrained from embarrassing her with this question. Instead, I proceeded to carefully sidestep the burden of making the decision for her. I suggested that one of the two graduate programs was more aligned with her interest in journalism and the other program was more aligned with her interest in information technology. In essence, I attempted to put the decision back in her lap. Alas, I did not receive a reply, despite the fact that I had taken some time to craft my response. A participant in a recent seminar told me that I had most likely disappointed—even failed—her in not giving her the answer she requested; the result was a loss of face that could not be overcome with a reply. I suspect that is the case.

Not being able to determine the cultural identity of the writer beyond what is suggested by her name, we can nonetheless determine the cultural characteristics of her residence in Singapore. Comparing the values assigned to Singapore for each of Hofstede's cultural dimensions with the values assigned to China and the United States, we see that the values for Singapore are very close to or the same as those for China on power distance and individualism. Masculinity and long-term orientation are near the middle of the range. However, Singapore scores the *lowest* of the 74 countries ranked on the dimension of uncertainty avoidance. This extremely low ranking may suggest that Usha is comfortable taking the risk of writing to a stranger for advice. However, her surname indicates that she is Chinese, so it is not possible to determine which set of cultural values is influencing her more strongly.

Cultural Dimensions[a]	United States	China	Singapore
Power Distance (104)	40	80	74
Individualism (91)	91	20	20
Uncertainty Avoidance (104)	46	30	8
Masculinity (95)	62	66	48
Long-term Orientation (118)	29	118	48

[a]The highest score attained by any country for each dimension is indicated in parentheses.

Example 3: E-Mail Letter From Indian Technical Writer

This next example is an e-mail letter from an Indian technical communicator whom I met briefly several years ago when giving a seminar on usability testing in Pune. This person has subsequently written to me on several occasions, most likely in an effort to establish a relationship that could be counted on in the future, when needed. As in the earlier example from Singapore, in this example I am being asked to influence the future direction of this person's career.

Dear Dr. Barnum

This is Anu Vishnu [name changed] from Pune, India. Just to recall our earlier communication, I was a privileged attendee at one of your "Usability Testing" workshops conducted at Pune in the year 2001. I had also written to you on an earlier occasion regarding the feasibility of usability testing in a banking scenario.

I write to seek your advice yet again, this time in the highly competitive field of Technical Writing. Doctor, I have moved from my earlier realm to Content Writing and taken a challenging assignment with a Japanese firm as a Technical Writer. Before I plunge headlong into this new profile, I wish to enhance my knowledge in the field. Besides reading up on the subject, I thought a few words of advice from an experienced and encouraging mentor like you would definitely help me in my new career.

Dr. Barnum, could you please guide me as to what books I can read up to refresh my knowledge in Technical Writing? Then, are there any particular kind of workshops/seminars (besides the regular Technical Writing / Communication ones) that I need to attend? I would like to gradually expand my horizons and move full time into Usability Testing. And needless to say, I would be greatly obliged to stay under your mentorship.

I have identified the following books as a startup. Kindly advise if I have got it right till now.

1. The Complete Idiot's Guide to Technical Writing
2. Technical Writing for Dummies
3. Technical Writing Basics: A Guide to Style and Form
4. Handbook of Technical Writing
5. The Chicago Manual of Style
6. The Microsoft Manual of Style for Technical Publications
7. The User Manual Manual
8. Untechnical Writing—How to Write About Technical Subjects and Products So Anyone Can Understand

I would also like to attend more of your interactive sessions. Are you likely to conduct any more of your speciality workshops in India? It would be a great pleasure and learning experience to interact with you Doctor.

This letter is a bit different from the other samples in that it states its purpose more directly and sooner than in the other examples. This may partly be because the correspondent works as a technical writer for a multinational company doing business globally, and also because the direct approach in correspondence is more likely to be taught in Indian education at the better schools (leading to a university education), based on Western traditions, particularly modeling British communication style. Despite the more direct approach used in this letter, the similarity to the other letters is in the emphasis on the relationship and the implied responsibility indicated by her request that I help her as "an experienced and encouraging mentor." She may be making the request with some earned degree of confidence because she has written to me on several prior occasions and always gotten a reply. The requirement of a reply is inferred from the message, indicating that I have a lot of face and that she has less face but that she strives to learn and grow, thereby gaining face. Her request for "a few words of advice" is politely understated, as there is no way to adequately respond to the request with a few words.

The tone of the letter is formal and the vocabulary sophisticated, although also employing some clichés, such as "plunge headlong" and "expand my horizons," like the letter from Singapore. She addresses me with the use of formal honorifics—Dr. Barnum and Doctor—and, in deference to my high status, states that she "would be greatly obliged to stay under [my] mentorship," implying again that I have an obligation to continue in this capacity. She closes with a flattering statement about the prospect of attending another of my workshops, as "it would be a great pleasure and learning experience to interact with you Doctor."

Comparing the values for Hofstede's cultural dimensions for India to those for China and the United States, we see that India's scores place it between China and the United States in all cultural dimensions except masculinity. These comparative scores support the similarities and differences evident in the Indian letter when it is compared to the two examples from China and Singapore discussed above.

Cultural Dimensions[a]	United States	China	India
Power Distance (104)	40	80	77
Individualism (91)	91	20	48
Uncertainty Avoidance (104)	46	30	40
Masculinity (95)	62	66	56
Long-term Orientation (118)	29	118	61

[a]The highest score attained by any country for each dimension is indicated in parentheses.

Following my detailed response with suggestions for resources inside India, as well as Internet resources, I received a response with just a few words of thanks. I cannot tell from the short reply if I failed to satisfy the request by not going through the list of proposed books and making specific recommendations on each one, as well as proposing others. Or perhaps the short reply was nothing more than what we would typically expect from a U.S.-based reply along the lines of "Thanks for getting back to me."

In the next example, a letter located on a Web site, I will be examining a somewhat different CMC, but one that still complies with the format and requirements of a business letter of introduction, as is the case in the examples already discussed.

Example 4: Chinese Letter to Web Site Visitor

A Chinese Web site seeking English-speaking customers for its export business and consultancy posts a letter to site visitors (Linan Euro-China Co., 2005) as a means of introduction and presentation of credentials. The target receiver is someone seeking an agent inside China for commercial transactions. We would assume that the intended audience includes those who can read English as well as those whose first language is English. This discussion of the cultural elements reflected in the letter and the potential problems that may result for readers is based on considerations for Western/North American readers.

Letter to Our site visitor

Dear Madam/Sir,

Hi, Welcome to Our website!

The primary purpose of Our website is to serve any foreign buyer, importer who is interested in purchasing any product from China, as We believe our quality Chinese products with the most competitive price will be very interesting and attractive to any foreign buyer and importer who are interested in purchasing from China for their better sales revenue and long term sustained business growth as well.

However, ***your cultural inexperience and language obstacle, your lack of knowledge and information on our Chinese exporting business operational procedure, relevant regulation, as well as our unique and typical local Chinese way of doing business, etc.*** . . . , all of these [pose] barriers and problems, We believe you will need to smooth them away before you can eventually succeed in your purchasing business from China. So, that is why We wanted to set up this website for you to help and assist you to do a successful purchasing business from China.

First, as an professional experienced Chinese businessman, trader and agent, We have been engaged in the field of exporting business from China to the international market for 12 years. We think We can help you to get a full clear picture and native insight into our overall exporting business, of which We believe will be very helpful for you to purchase and import properly and gain a right purchase channel from China.

Secondly, you will not have any language obstacle as long as you can communicate with us in English, although We are a Chinese, We can speak and write English fluently to maintain an effective direct business communication with you, please just always feel free to contact us at your early convenience.

We are listening to any inquiry, question, comment, suggestion and proposal, etc., comes from you, and We will get back to you as soon as possible.

As We are closely associated with many of our Chinese export oriented manufacturers, We can get the product you need and provide a very competitive offer to meet your purchase business need and market demand, as We have been working on multiple export products ranged from **Clothing products and construction products to electric products**. you may visit our website for more information.

Anyway, We are sure that you will need a capable agent who can assist your business in China, if so, please go to check out our agency service profile. We believe it is better for you to have a reliable native helping hand to assist your business in China than to go alone at your own risk, which sometimes could be just the "hit-and-miss."

We sincerely hope Our agency service will be interesting and helpful for your business in China, We are looking forward to establishing a lasting long term business relationship with any bona fides serious overseas business and client for the mutual benefit, please feel free to contact us at your early convenience.

Yours Sincerely,
Mr. Wang Hua

Like the letter from the Foreign Affairs Office of the Chinese university, this letter opens with an informal greeting to establish a friendly tone. Then, in similar fashion, it begins the presentation of its message by emphasizing the mutual benefits of establishing "long-term sustained business growth." Numerous instances of the use of the pronoun "We," always capitalized, perhaps for emphasis, suggest that the relationship is being stressed. At the same time, the message emphasizes the expertise one would expect to receive from this Chinese partner, noting (in bold italics) that "your cultural inexperience and language obstacle, your lack of knowledge and information on our Chinese exporting business operational procedure, relevant regulation, as well as our unique and typical local Chinese way of doing business, etc." put the foreign customer at a disadvantage that can be addressed via the superior skills of the Chinese partner. This reference to the superior expertise of the Chinese partner continues in the following paragraphs.

Setting aside issues of translation or other second-language problems in the letter, there is no mistaking its tone or intent. The Chinese businessperson is most likely speaking from a cultural position that recognizes higher authority and expertise in Confucian culture as the basis for such a partnership, and blunt references to "your" deficiencies reflect this position. The resulting dependency of the non-Chinese potential partner on the expertise of the Chinese partner is reinforced near the end of the letter with what might be perceived by the Western reader as a hint of the adverse consequences that could result from not accepting the expertise of the Chinese partner, while at the same time the letter emphasizes the advantage resulting from "a helping hand" in the statement: "We believe it is better for you to have a reliable native helping hand to assist your business in China than to go alone at your own risk, which sometimes could be just the 'hit-and-miss.'" The letter ends by again emphasizing the opportunity to establish "a lasting long term business relationship" for the "mutual benefit" of both parties. Not only does this ending statement reinforce the strong long-term orientation of the writer, but it also suggests the mutual face-giving results that can be derived from such a relationship. As China is a high-context, low-individualism, high-long-term-orientation culture, this approach makes sense from the writer's cultural perspective. However, potential Western business partners may find the tone of the letter off-putting.

IMPLICATIONS FOR TEACHING

Cross-cultural communication has gained enough of a toehold in the American academic curriculum that there are courses on the subject, as well as, typically, a chapter or section on it in business and technical communication textbooks. Recent articles and collections of essays have addressed the impact of technology on communication. For example, a special issue of the *Journal of Business and Technical Communication* on "Communication Challenges from New Technology" includes not only the earlier-cited article on e-mail overload but another article cleverly entitled "Ari, R U There? Reorienting Business Communication for a Technological Era" (Reinsch & Turner, 2006). Among the observations and pronouncements of the authors, this one is particularly notable: "Changes in communication technologies and the effects of those changes on tasks, jobs, organizations, and lifestyles (along with increasing globalization and the consequent necessity of interacting across cultural and linguistic boundaries) should make communication breakdowns—and communication—more visible to practitioners" (p. 349). Their solution to the growing number of breakdowns is for educators to focus even more on rhetorical principles—the Ari of the title is instant-messaging shorthand for Aristotle—particularly those of audience and situation.

Herrington and Tretyakov (2005) arrive at the same conclusion in their chapter in the edited collection *Online Education: Global Questions, Local Answers* (Cook & Grant-Davie, 2005). Sharing their experience of teaching a cross-cultural communication unit with American and Russian students in a distance-learning environment, Herrington and Tretyakov emphasize that "technical communication cannot be taught by rote, cannot be taught as a system, and cannot be taught as a cleanly packaged structure. The context of technical communication changes instance by instance, and that affects its function, so students can only learn how to adapt to it" (p. 281). Describing the process of teaching in an experiential learning environment as "chaotic and confusing" (p. 282), they emphasize the importance of recognizing the impossibility of making confident pronouncements about any cross-cultural communication exchange.

That is not to say that we should give up the effort. Composition and English as a second language instructors have been analyzing the rhetorical issues raised by student writing in cultural contexts, launching the study of contrastive rhetoric with Kaplan's coining of the phrase in 1966. In *Contrastive Rhetoric Revisited and Redefined*, Kristin Woolever (2001) contributes a chapter that specifically addresses the issue for the business and technical communicator:

> Electronic wizardry can provide the opportunities for communication, but only a thorough understanding of the rhetorical contrasts among cultures will allow that communication to be productive. Professionals working in business and industry need to transform their rhetoric to accommodate the multicultural traditions and expectations of people they may never speak to face-to-face. (p. 49)

Woolever calls for "pedagogical reform in higher education to better reflect the actual needs of global industry and to better prepare students to enter these professions" (p. 49). Certainly, those of us who teach in postsecondary education are well aware of the need and are attempting to meet the challenge. Using our texts' coverage of cross-cultural communication, we take our students through various exercises to increase their awareness and make them more culturally sensitive.

But the task is complex and the need grows greater. Woolever is correct in predicting that the problem is far from solved, and that a transformation in thinking and in teaching is needed. She writes: "As the leader in developing programs for technical communication, the United States is well positioned to lead this transformation. But until we can coherently connect the needs of the work world with the practices in our classrooms, we will continue to do a disservice to our students and jeopardize our position as a world leader in the field of technical communication" (p. 63). This means that we, as technical and business communication instructors, must become lifelong learners in order to teach this evolving and increasingly important area of cross-cultural communication.

How to do this? We have good teachers among us, both in academe and in business, with some of the authors of this volume surely among this small but growing group. We can read what they have written, along with the growing collection of articles and books on international and intercultural communication. Some of these are generalized; others are country/culture specific. Other steps we can take include finding inventive ways to partner with business to learn how it is overcoming the rhetorical gaps that are evident in a great deal of cross-cultural communication. We can also partner with colleagues in other cultures to share the interpretation and analysis of common forms of communication, so as to learn from and teach each other. We can expose our own interpretations to cultural experts in the target culture so as to gain an understanding of how even our best efforts are being interpreted. With these and other similar efforts, we can better prepare ourselves to prepare our students for the issues they will commonly face in their professional lives as they become subject to and creators of content for computer-mediated communication.

REFERENCES

Archee, R. K. (2003, September/October). Online intercultural communication. *Intercom,* 40–41.

Barnum, C. M., & Li, H. (2006). Chinese and American technical communication: A cross-cultural comparison of differences. *Technical Communication, 53*(2), 143–166.

Bovée, C. L., & Thill, J. V. (2005). *Business communication today* (8th ed.). Upper Saddle River, NJ: Prentice Hall.

Callahan, E. (2005). Cultural similarities and differences in the design of university Web sites. *Journal of Computer-Mediated Communication, 11*(1) article 12. Retrieved February 21, 2007, from http://jcmc.indiana.edu/vol11/issue1/callahan.html

Campbell, C. P. (1998). Beyond language: Cultural predispositions in business correspondence. *Proceedings of Region 5 STC Conference, February 5, 1998.* Fort Worth, TX.

Cardon, P. W., & Scott, J. C. (2003). Chinese business face: Communication behaviors and teaching approaches. *Business Communication Quarterly, 66,* 9–22.

Connor, U. (1996). *Contrastive rhetoric: Cross-cultural aspects of second-language writing.* New York: Cambridge University Press.

Connor, U. (2002). New directions in contrastive rhetoric. *TESOL Quarterly, 36,* 493–510.

Cook, K. C., & Grant-Davie, K. (Eds.). (2005). *Online education: Global questions, local answers.* Amityville, NY: Baywood.

Duan, P., & Gu, W. (2005). Technical communication and English for special purposes: The development of technical communication in Chinese universities. *Technical Communication, 52*(4), 434–448.

Durham, M. (2003). Language choice on a Swiss mailing list. *Journal of Computer-Mediated Communication, 9*(1). Retrieved February 21, 2007, from http://jcmc.indiana.edu/vol9/issue1/durham.html

Ess, C., & Sudweeks, F. (2005). Culture and computer-mediated communication: Toward new understandings. *Journal of Computer-Mediated Communication, 11*(1) article 9. Retrieved February 21, 2007, from http://jcmc.indiana.edu/vol11/issue1/ess.html

Faiola, A., & Matei, S. A. (2005). Cultural cognitive style and Web design: Beyond a behavioral inquiry into computer-mediated communication. *Journal of Computer-Mediated Communication, 11*(1). Retrieved February 21, 2007, from http://jcmc.indiana.edu/vol11/issue1/faiola.html

Fry, J. (2007, April 2). Stop the e-mail madness! *Wall Street Journal Online*. Retrieved April 3, 2007, from http://online.wsj.com/public/article_print/SB117519357344553426.html

Guffy, M. E. (2003). *Business communication: Process and product* (4th ed.). Mason, OH: South-Western.

Hagan, P. (1998). Teaching American business writing in Russia: Cross-cultures/cross-purposes. *Journal of Business and Technical Communication 12*(1), 109–126.

Hall, E. T. (1989). *Beyond culture*. (2nd ed.). New York: Anchor.

Hermeking, M. (2005). Culture and Internet consumption: Contributions from cross-cultural marketing and advertising research. *Journal of Computer-Mediated Communication, 11*(1) article 10. Retrieved February 21, 2007, from http://jcmc.indiana.edu/vol11/ issue1/hermeking.html

Herrington, T., & Tretyakov, Y. (2005). The global classroom project: Troublemaking and troubleshooting. In K. C. Cook & K. Grant-Davies (Eds.), *Online education: Global questions, local answers* (pp. 267–283). Amityville, NY: Baywood.

Hofstede, G., & Hofstede, J. (2005). *Cultures and organizations: Software of the mind* (Rev. 2nd ed.). New York: McGraw-Hill.

Houp, K. W., Pearsall, T., Tebeaux, E., & Dragga, S. (2002). *Reporting technical information* (10th ed.). New York: Oxford University Press.

ITFacts. (2007). E-mail usage. Retrieved May 30, 2007, from http://www.itfacts.biz/index.php?id=C0_8_1

Kaplan, R. B. (1966). Cultural thought patterns in intercultural education. *Language Learning, 16,* 1–20.

Lee, O. (2000). The role of cultural protocol in media choice in a Confucian virtual workplace. *IEEE Transactions on Professional Communication, 43*(2), 196–200.

Lee, W., Tan, T. M. K., & Hameed, S. S. (2005). Polychronicity, the Internet, and the mass media: A Singapore study. *Journal of Computer-Mediated Communication, 11*(1) article 14. Retrieved February 21, 2007, from http://jcmc.indiana.edu/vol11/issue1/wplee.html

Linan Euro-China Co., Ltd. *Welcome to China Linan City*. Retrieved August 6, 2006, from http://www.linanwindow.com

Ma, R. (1996). Computer-mediated conversations as a new dimension of intercultural communication between East Asian and North American college students. In S. C. Herring (Ed.), *Computer-mediated communication: Linguistic, social and cross-cultural perspectives* (pp. 173–186). Philadelphia: John Benjamins.

Nisbett, R. E. (2003). *The Geography of thought: How Asians and Westerners think differently . . . and why.* New York: Free Press.

Reinsch, N. L., Jr., & Turner, J. W. (2006). Ari, R U there? Reorienting business communication for a technological era. *Journal of Business and Technical Communication, 20,* 339–356.

Sandberg, J. (2006, September 26). Employees forsake dreaded e-mail for the beloved phone,. *Wall Street Journal Online.* Retrieved September 26, 2006, from http://online.wsj.com/public/article_print/SB115923906128273954.html

Shipley, D., & Schwalbe, W. (2007). *Send: The essential guide to e-mail for office and home.* New York: Knopf.

Singh, N., & Pereira, A. (2005). *The culturally customized Web site: Customizing Web sites for the global marketplace.* Oxford, England: Elsevier Butterworth-Heinemann.

Thomas, G. F., King, C. L., Baroni, B., Cook, L., Keitelman, M., Miller, S., et al. (2006). Reconceptualizing e-mail overload. *Journal of Business and Technical Communication, 20*(3), 252–287.

Woolever, K. R. (2001). Doing global business in the information age: Rhetorical contrasts in the business and technical professions. In C. G. Panetta (Ed.), *Contrastive rhetoric revisited and redefined* (pp. 47–64). Mahwah, NJ: Erlbaum.

Yum, Y.-O., & Hara, K. (2005). Computer-mediated relationship development: A cross-cultural comparison. *Journal of Computer-Mediated Communication, 11*(1) article 7. Retrieved February 21, 2007, from http://jcmc.indiana.edu/vol11/issue1/yum.html

Yunker, J. (2003). *Beyond borders: Web globalization strategies*. Berkeley, CA: New Riders.

Zhu, P., & St.Amant, K. (2007). Taking traditional Chinese medicine international and online: An examination of the cultural rhetorical factors affecting American perceptions of Chinese-created Web sites, *Technical Communication, 54*(2), 171–186.

CHAPTER 7

Meeting Each Other Online: Corpus-Based Insights on Preparing Professional Writers for International Settings

Boyd Davis, Tsui-ping Chen, Hui-fang Peng, and Paul Blewchamp

CHAPTER OVERVIEW

International online interaction raises issues of language, culture, and the negotiation of common understandings about collaborative tasks. As one of the first events among members of virtual teams, self-introduction is important in terms of first impressions. The analysis of a small corpus of online Chinese-U.S. student letters of self-introduction identifies subtle usage features that index cultural habits and expectations that are important for the pedagogy of professional writing in international settings.

Faculty preparing writers in and for international collaboration around writing find themselves entangled in new kinds of questions about online interaction and its contexts. As Wilson and Peterson (2002, p. 459) comment in a study of online communities, "Using English-language data such as emails, chat room transcriptions, and bulletin board posts, Crystal (2001) asserts that new varieties of language are indeed emerging from new technologies, but suggests that cultural and linguistic differences which could influence online interactions remain under-researched" (cf. Dauterman, 2005, St.Amant, 2002). The authors of the present chapter are all university faculty members in linguistics, who are charged with teaching writing to students anticipating international careers: Two of the authors are Chinese, one is British, one is American, and all four at the time of the research for this discussion were teaching in Taiwan (Republic of China). Our discussion of cross-cultural writing in international settings grows from our examination of a very small corpus, outlined below in Table 1.

The first component of the corpus was a two-way exchange of e-mail between Chinese undergraduate students who were business majors and U.S. undergraduate and graduate students from a variety of majors, taking a course titled Language, Health and Aging online with the American visiting professor, via the CENTRA platform, while she was in Taiwan. We asked our U.S. and Chinese students to participate in a brief online exchange, hoping to identify some of the cross-cultural issues in self-introductions, since that could affect team formation in professional

Table 1. Two Components of a Small Corpus of Student Writings

	E-mail	Midterm
Asynchronous setting	Two-way, two-time exchange, U.S.-Taiwan students.	One-way, one-time attachments from U.S. students.
Participants (interpersonal)	Chinese student-U.S. student direct exchange, with copies to instructors.	31 U.S. students to Taiwan teachers via U.S. professor teaching online from Taiwan.
Details	31 Americans (6 male, 25 female) e-mail self-introductions to unknown audience. Length varies. 23 Chinese (4 male, 19 female) e-mail self-introduction to unknown audience **before** reading U.S. e-mail. Length varies. 22 Chinese (1 male, 21 female) e-mailing self-introduction in reply to Americans **after** reading U.S. e-mail self-introduction. Length varies. 1 pair (Chinese male, U.S. female) continues for 4 week exchange.	31 Americans (6 male, 25 female) write offline and e-mail as attachment, 10 extended definitions of key terms, set in the context of the writer's profession or discipline.

settings. The second component was a set of U.S. midterm examination essays defining terms, written offline and emailed to us, so that we could look for genre features. We hoped this small corpus would help us identify habits and expectations of both groups of student writers with regard to self-presentation and informational writing. Our discussion will focus primarily on one of these components, the exchange of self-introductions and responses, since self-introductions are typically one of the first events to take place among members of virtual teams and are important in terms of first impressions.

To protect confidentiality, students' consent to the use of their writings for instructor analysis was obtained, and the writings were anonymized and aggregated as data; personal names were changed; identifying details were deleted or changed; and we made the standard offer of deleting writings by any student who requested it (none did).

LITERATURE REVIEW

Our use of e-mail letters of self-introduction to look at features of culture and language in online contexts required us to explore the research in several areas: current linguistic pedagogy for second-language writing; international business communication; and constructs of culture in technical writing discussions of international online communications. With regard to the e-mail letters of self-introduction, we highlight research on self-presentation and then raise the issue of what variety of English one might expect and how one might expect it to be taught: from a first-language, second-language, foreign-language, additional language, or specific-purposes perspective. Seasoned technical/professional writers know they should expect subtle yet important lexical and rhetorical differences between them, keyed to language variety. These differences can cause problems when editing the diction and syntax of the final products. The differences cause even more problems if they go unrecognized, for example, when readers of e-mail try to decide if what they are reading includes a question, and if so, whether or how to reply to it.

Second-Language Writing Pedagogy: International and Online Contexts

An international cross-cultural research perspective that draws on insights from current pedagogy and practices in the teaching of English writing to second-language learners can be exemplified by studies such as those of Daoud (1998) and Wong (2005) and adapted to an online context. Daoud discusses the way in which students learn new rhetorical structures through a cross-cultural exchange of student essays. Daoud's emphasis on the exchange of preparatory questions and outlines prior to each group's writing for the other's eyes is one that instructors using Internet-based communications can easily incorporate. Wong (2005) discusses the kinds of mental representations that second-language writers of English use when composing; his use of the talk-aloud protocol can be further adapted to studies of composing online. Li and Flowerdew (2007) identify the impact of the "shaper," the formal and informal editors of scientific research articles by nonnative-speaker scholars, calling for "language professional-scientist" partnerships (p. 114). Technical writers whose background includes some study of language analysis and cultural constructs can provide the kinds of professional editorial services needed by subject professionals.

A useful starting point for researchers and teachers wanting to use Internet-based communication for cross-cultural collaboration is Warschauer and colleagues, both of whose two books published in 2000 offer still-standard discussions and suggestions for Internet- and network-based teaching of English. The recently published collection, *The Multilingual Internet*, edited by Danet and Herring (2007), focuses on computer-mediated communication in multiple languages, including nonnative varieties of English: It provides a superb introduction for technical writing students across the globe to their future audiences.

International Business Communication

Discussions of intercultural features in international business communication do not always include Internet-based communication or technical writing as a focus; nonetheless, they reflect current scholarly and pedagogical preoccupations, and these features could be useful for teaching technical writing, since online communication blends oral and written features of language (Davis & Brewer, 1997). In examining computer-mediated conversation between Asian and American college students, Ma (1996) explained that the two groups of students had different views about directness and self-disclosure. Xiao and Petraki (2007), for example, surveyed and interviewed Mainland Chinese students studying in Australia to elicit their intercultural communication competence as well as their perceptions about intercultural communication. Xiao and Petraki noted that the Chinese students were well aware that book-knowledge of a language did not prepare them for intercultural business collaboration or team building, which suggests a readiness to collaborate. However, collaboration may occur in a different fashion. Ortiz (2005) assessed rhetorical and linguistic features of written business communication on the U.S.-Mexico border, looking specifically for features associated with Mexican business discourse "such as indirectness about purpose, ornateness and fluidity in style, and placing higher emphasis on personal rather than business issues" (p. 28). Ortiz noted considerable change over the last 10 years in the way in which

Mexican border professional writers create text: They have now accommodated to American writing conventions. That may or may not be desirable, depending on the audience.

Technical Writing and Culture: A Call for Change

A number of scholars have recently focused on issues specific to the field of technical-professional writing and the way that field is handling issues of culture in online contexts. Niemeier and coeditors (1998) presented a cornucopia of fine discussions of cultural context in business communications including Campbell's (1998) review of high- and low-context cultures. This is part of a larger discussion across the last four decades that owes much to seminal work on cultural domains by Edward T. Hall, on cross-cultural rhetoric by Robert Kaplan, and on cross-cultural business values by Geert Hofstede. Each of these has been gently contested, for example, by Connor (1996) on rhetoric, and in a recent discussion by Kumaravadivelu (2003). Over the years, Kaplan has modified his earliest pronouncements, but many researchers continue to find his initial concepts good launching pads. An outstanding discussion is provided by Artemeva (1998), who combines genre (the periodic engineering report) with cultural perspectives, Russian and Canadian, to make the case that the technical writing consultant is a cultural interpreter. Looking to future training, Woolever (2001) bases her comparison of traditional and "forward-looking" teaching practices (p. 63) on the need for a teaching reform that will support the need in global business for virtual teams. The features characterizing forward-looking practices are subtle and far-reaching, such as assuming "that students will write for many different cultures" (p. 63). This expanded audience raises a number of issues including translatability.

We believe that some kind of concretization is needed that will allow students to see what "write for many different cultures" means, and that will welcome Asian teacher-researchers to the table. This belief is shared by others. Giammona (2004) first reviews global trends in technical and professional writing and then calls for a repackaging of training that goes beyond outsourcing. Liaw (2007, p. 224) focuses on computer-based pedagogy, in discussing her use of "computer technology to assist L2 learners' growth in either linguistic or intercultural competence." She reports that networked online materials and online interactions increased Chinese students' proficiency and fluency in writing and demonstrated their new awareness of cultural aspects of their interactions.

Our microanalysis of online letters of self-introduction and midterm definitions grew out of a desire to pin down features of language in computer-supported contexts that could be built into teaching and training that prepares people for collaborative international projects.

Self-Presentation

We use corpus analysis (Biber, 1988; Sinclair, 2004) to identify features of self-presentation in both the letters and the midterm exams, although in this discussion we focus on the letters. Both of these writing tasks invoke reader and writer expectations for interpretive reception, and may infringe on assumptions about politeness and face for self and other (Lee & Swales 2006; Scollon & Scollon 2000). Specific phrasal means of self-identification may have different cultural reverberations. The *I* in letters of introduction might be seen as an autobiographical or as a discoursal construction projecting an identity (Tang & John, 1999). Tang and John (1999, p. S25) claim that a writer may play roles at any of three levels: societal ("father," "Malaysian"), discourse ("client," "student"), and genre (the "architect of the essay"). Choosing one or another of these roles for self-presentation could affect the reader's reception of a letter of introduction. We will return to self-presentation and expectation as we reference ways in which the presence or absence of small talk—face-to-face or online—apparently affects socialization among people who will form virtual as well as face-to-face teams.

It is useful to look closely at features of self-introductions. One might assume that, given the ubiquity of self-introductions, people know what a letter of self-introduction will include. However, our assumptions about words and usage, or the presence of larger "moves" such as greetings and closings were confounded by what writers were actually doing. Having one's expectations confounded is not unusual, as described in Kirkpatrick's (1991) early comparison of Chinese and English information sequencing in letters of request, and more recently in a study of internal corporate e-mail in English as lingua franca by Kankaanranta (2006). The latter includes an examination of framing moves for salutation, closing, and signature in the e-mails across the company's communications genres. Kankaanranta adds that while learning a company's preferred conventions for communicative interaction takes place on the job, case studies can be used to give students a needed foundation.

Preliminary analysis of our own small corpus helped us to identify features with the potential to influence online interactions among our students and among ourselves that can be used as case studies for teaching purposes. Such case studies may enable us to help writers in international settings edge out of their comfort zone, initiate and sustain exchanges, and find common ground for public as well as private writing.

Choosing a Perspective on English for Online Exchange

There is at least one other area of research that must be highlighted: research on the level and version of English that is to be chosen for online collaboration. In any area of instruction in academic or occupational English, it is necessary to consider how best to develop an international collaboration that fosters ongoing discussion of multiple perspectives on technical communication for international online environments, perspectives that are coequal in value, and subject to negotiation by all parties. In the background of this discussion are the issues cogently argued by British scholar David Graddol in *English Next* (2006). These issues, which include the reduced prestige of native speakers and the reviewing of attitudes to nonnative varieties of English, deserve consideration by anyone planning to teach English, teach English writing, or teach writers to develop expertise in various genres of written English, however defined.

Many of the discussions of teaching professional, technical, or discipline-specific writing over the last two decades have focused on teaching academic English to international students or workers in countries whose first or dominant language is English. However, researchers and instructors have been eager to learn about multiple cultural expectations for writing conventions, and to incorporate them in instruction. For example, Angelova and Riazantseva (1999) offer a case study of international graduate students that positions academic writing as part of a necessary "disciplinary enculturation" (p. 491), and offers specific advice for instructors of written communication in English. More recently, Coverdale-Jones (2006) has summarized a special journal issue on international (Chinese) learners at British universities by noting a dilemma for instructors in English-as-a-first-language countries: They "wish to avoid 'otherisation' of the Chinese learner, but find that we need to develop strategies for dealing with the numerous students when they arrive" (p. 148).

In the last few years, however, research and publications based on work in locations where English is a foreign language (FL) have become "clearly dominant" (Silva & Brice, 2004, pp. 71–72): The scholarship is being carried out all over the world, and the genres of writing being taught have moved from an emphasis on the general or school essay to "real-world" contexts. Silva and Brice note a proliferation of research on FL (not second-language, or SL) writing, which is seen as "evidence both of the growth of L2 writing as a discipline and the changing role of writing in the 21st century" (p. 79) and of an expanded number of programs, courses, and writing centers around the world.

Nickerson (2005) identifies two examples of researcher-practitioner reports of praxis in the teaching of English for specific business purposes, one on surveys and the other on case study methods, each of which can be adapted to the concerns of first- and second-language speakers alike. Nickerson notes a move among researchers to "shift from a focus on language proficiency . . . to a concern with teaching the rhetorical strategies . . . that have [been] identified as effective strategies through the investigation of various business genres" (p. 374). Since genres of communication have associated rhetorical strategies, we found that looking closely at language use in a particular genre, here, e-letters, gave us insight into what and how we might teach novice writers.

EXAMINING LETTERS FOR CULTURAL FEATURES: A RESEARCH CONTEXT FOR USING CORPUS ANALYSIS

Corpus-based studies and the development of publicly available as well as privately compiled corpora of business letters have become a part of scholarship in technical writing, as exemplified by Kessapidu (1997), Upton and Connor (2001), Yasumasa Someya's Learner Business Letters Corpus (2000), and business and professional texts such as the Wolverhampton Business English Corpus (2001) and the in-progress Corpus of Professional English (2004-present). Lee and Swales (2006) discuss the usefulness as well as the limitations of the self-compiled corpus for pedagogical purposes. Recent corpus-based studies not focused on English for specific purposes (ESP) or English for academic purposes (EAP) may be exemplified by Kaufer and Ishizaki (2006), who report on "canned" letters; Lavid and Taboada (2004), who discuss multilingual administrative forms; and Orr (2006), who offers an overview of the evolution and contribution of corpus-based analysis to professional and technical writers.

E-mail exchanges across classes, continents, and cultures have grown steadily more sophisticated over the last 20 years. Greenfield (2003, p. 46), for example, is able to refer to an e-mail project as one that exemplifies "widely accepted theories and methods for modern second language instruction: cooperative learning, communicative language learning, process writing, project-based learning, and an integrated approach" in a discussion of Hong Kong writers paired with American writers for the purpose of improving English skills. Matsuda, Canagarajah, Harklau, Hyland, and Warschauer (2003, p. 164) summarize a decade of research on e-mail exchanges, citing Warschauer (1999) on how exchanges failed "when they [were] added on in a mechanical fashion, and thus, were not viewed as meaningful or significant."

Many instructors who write about developing cross-cultural e-mail exchanges for their students suggest that a minimum number of interactions should be stipulated and that the exchange should be an important part of the course of study, including the course grade (see, for example, several discussions included in Warschauer & Kern, 2000). We have used such pedagogy in the past (Davis & Thiede, 2000, 2006). For this project we chose, instead, with the cooperation of our students in the United States and in Taiwan, to put together a small, two-component corpus that did not make such stipulations. The brief exchange of self-introductions and replies between Chinese and American students was the equivalent of a wave from passing but friendly strangers who had been asked to demonstrate greeting behaviors. In brief, we wanted the e-letters to replicate the often awkward start-up of in-class and virtual teams, to see what might help smooth the path for students who will eventually be involved in cross-cultural work on Internet-supported teams.

METHODOLOGY: TEASING OUT FEATURES OF LETTERS OF SELF-INTRODUCTION USING A CORPUS APPROACH

In this section, we first review rhetorical moves setting up politeness strategies as discussed by Upton and Connor (2001), in order to ground our discussion of politeness strategies in the letters from Chinese students. Our analysis, using WMatrix, an online corpus-analysis tool, identified several areas of sensitivity that a teacher of technical writing could share with students: self- and other-reference; politeness phrases; question types that may or may not invite responses (see also Davis & Mason, 2006, on the use of questions in electronic discourse); and openings and closings. These four areas represent grammatical, lexical, and discourse levels of usage and can be used by international technical writing teachers to demonstrate that all three levels are involved in international cross-cultural writing.

Upton and Connor (2001) illustrate the usefulness of small, genre-specific corpora in their examination of letters of application by Finns, Belgians, and Americans from the Indianapolis Business Learner Corpus. They identify seven rhetorical moves, from an explanation of where the applicant learned of the opportunity to the attachment of the resume (p. 318). Move 4 is an *invitation* in the sense that it can "Indicate desire for an interview or a desire for further contact, or specify means of further communication/how to be contacted." Upton and Connor examine the modals *could, may, shall, might, should* and *would*, for their social-interactional or social-pragmatic impact and find that their use correlates with the presence of this particular move, and with the use of formulaic expression including multi-word expressions (MWEs), which differed across the three groups. Though differing with regard to the way in which they used these modals and this move, both Belgians and Americans used them more than the Finns. As Upton and Connor note, politeness strategies as signaled by specific tokens such as modals used to qualify an expression or intent, *would you be willing to consider*, will differ not only by "writer concepts of how politeness is expressed" but also by "whether that addressee is American, Belgian, or Finnish" (2001, p. 325).

Our first objective was to locate features that could impact interaction in even the most minimal online exchange, in letters of self-introduction, which often initiate collaborative work. Which strategies would either group choose for self-presentation that might signal their notion of what a self-introduction should include? Here we were influenced by a study reported by Al-Khatib (2001) of the impact of audience on the composition of personal letters and an investigation by Belz (2005) of the impact of different types of intercultural questioning in Internet exchanges on participant perceptions and behaviors. Al-Khatib (2001) analyzed cultural and linguistic features in personal letters written in English by 120 Jordanian (Arabic-speaking) students to an unknown audience of hypothetical friends. Al-Khatib's review of English letter writing in such situations (p. 186) is particularly interesting, as it identifies characteristics of informal letters as personal writing: concrete words, short conversational sentences, and culture-specific conventions for greetings, closure, gender (same-sex address) and communal attributes such as invitations and offers, which signal friendliness and index politeness strategies (pp. 187–190). Accordingly, we compared the ways in which friendliness and politeness strategies were handled in letters from one set of Chinese students writing to an unknown audience and in letters from a second set of Chinese students writing to an audience whose members they had "met" through self-introductions.

Corpus analysis was conducted with WMATRIX (Rayson, 2007), an online corpus analysis tool that supports keyword in context (KWIC) concordancing, and semantic and part-of-speech tagging. Our second objective was to begin to identify features that might be gender cued (Mondorf, 2002) in writing letters; this was our focus here, and in the extended definitions of the midterm, a task that frequently occurs early on in collaborative writing work.

FINDINGS

Self- and Other-Reference

Figure 1 examines by percentage the most frequent parts of speech used by Chinese and American women, since there were so few men in the sample. Here, we looked especially at self-other reference, thinking it might be a pointer to finding in letters of self-introduction some trace of what Holmes (2006, p. 75) calls "relational practice" in workplace settings: "people-oriented behaviour which oils interpersonal wheels at work and thus facilitates the achievement of workplace objectives." Relational practice as Holmes defines it is clearly crucial to successful online international collaboration: building it in means finding the small details within the larger picture.

As shown in Figure 1, all three groups (American, Chinese writing with no audience in mind; Chinese writing in reply to an American model) used self-reference, though Chinese with no audience used it more frequently. None of the groups had high frequencies of third-person or relative pronouns in their letters. Chinese used the direct-address *you* with great frequency, unlike the Americans. On the basis of frequency of nominals, American letters seem to be more lexically dense, more like the professional letters seen in asynchronous threaded discussion than informal, conversational e-mail notes in their use of prepositions and articles (Biber 1988; Davis & Brewer 1997). When arranging self-introductions, then, it would be wise to provide a model or some discussion of tone or register so that writers can adjust their lexical registers for self-reference.

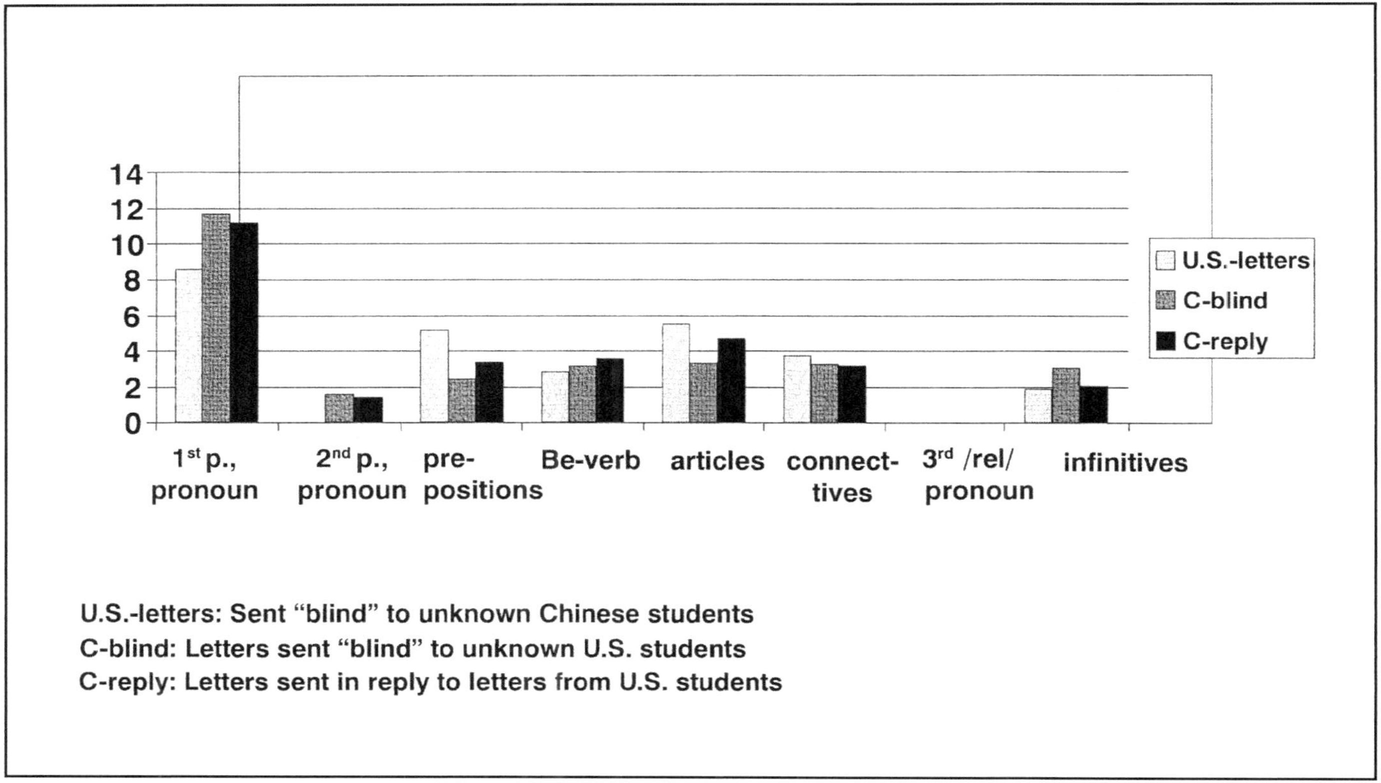

Figure 1. Most frequent features in letters by U.S. and Chinese women students.

Sensitivity to Politeness-Phrases and Modals

Not everyone has the same norms for directness or politeness. Belcher (2004) discusses differing cultural norms for directness in the German-American exchange she studied, noting that "Eric may interpret this directness as rudeness, whereas Anke and Catharina may opine that they are just being honest" (p. 26). American writers may have found Chinese expressions of directness disconcerting. These sentences from the e-letters are examples: "I hope to hear you soon and be friends with you"; "I wish you could realize me through the mail"; "Waiting for your letter in reply!" If readers of these letters were having a first encounter with Taiwanese writers, they might think the sentences presented a demand, a greater-than expected emphasis, rather than conventionally polite phrases, which could result in lowered motivation to reply. On the other hand, as with Belcher's Eric, the Chinese writers were not yet advanced either in fluency or in confidence: they could very well have missed or misunderstood cues of politeness in the message they were sending or receiving, such as modals used to qualify or soften direct requests.

Table 2 displays the modals and "polite phrases" used in the Chinese letters of self-introduction. Note that the Chinese business major students use three modals, *can, would*, and *may*, in their writings, although only *would* is used in phrases of courtesy. In contrast, U.S. writers present seven modals, *would, will, can, must, could, may,* and *should,* often using *would, could, will,* and *must* in courtesy phrases. Native speakers and new learners alike can benefit from reviewing how modals are used in different genres across online settings; insights and arrays of data are both obtainable from small corpora of authentic exchanges, as numerous scholars have remarked.

Table 2. Chinese Self-Introduction Letters: Use of Modals and "Polite Phrases"

	Count	Representation
Modals		
Can	34	Ability
May	4	Possibility
Multi-word expression (qualifying phrases including modal)		
Would like to	8	To show politeness
Look forward to	2	To show politeness

Note: Total number of word tokens in SCU letters: 4288

Sensitivity to Question Types

Belcher's (2004) analysis of questions supported our interest in why Chinese writers chose to respond to some questions posed by Americans but not others: Was it content, or was it the form of the question? Since most Taiwanese students receive their language education in instructional settings, they may be more consciously aware of the types of questions that are typically taught in school, that is, WH-questions (*What is that?*) and yes/no questions (*You really do like movies, don't you?*). The questions can be closed or open ended to some degree (*Why did you take that painting?*). In our corpus of messages, "Jo-Ellen" and "Molly" asked a total of 18 direct questions (both open ended and closed), and seven indirect questions. Examples of indirect questions include Jo-Ellen's comment to A: "It sounds like you are very busy teaching with all your part-time jobs"; however, A failed to say anything at all about her jobs in her response. Molly wrote to C: "It must be exciting to have your first full time job that is your field!" but no feedback was forthcoming from C, either. Students who have not been taught indirect question strategies in the target language may not recognize their illocutionary or perlocutionary force; they may not recognize invitations to disclose or realize that they are offering such invitations.

In all, 72% of the direct WH- and yes/no queries obtained a response, whereas indirect questions failed to get responses 57% of the time. Two additional aspects that surfaced warrant mentioning. First of all, the relatively high percentage (82%) of topic initiations introduced by questions in the Chinese letters of self-introduction might suggest some intention to keep contact with their English as a first language (L1) counterparts or prolong the interaction with them by making requests or asking other questions they hoped would appeal to their audience. Another observation seems to be gender related, keyed to content: Both of the L1 writers, Jo-Ellen and Molly, briefly mention their plans for marriage in their self-introduction, eliciting best wishes from three of our English as a second language (L2) writers, all but one of whom were female. Neither L1 writers mentioned their weddings when writing to each individual L2 reader; rather, weddings were mentioned relatively briefly in the earlier general introduction: "I am preparing to graduate . . . in May and I am getting married in June." As brief as the message was, female L2 writers picked up the message and responded to it.

Sensitivity to Opening/Closing the Exchange

When teaching students cross-cultural varieties of politeness, it is crucial to point out the differences existing in culturally based variations on speech acts. For example, in their report on compliment variation, Hondo and Goodman (2001) review current literature on compliments among speakers of five languages (American English, Chinese, Japanese, Egyptian Arabic, and Spanish), and they remind instructors to include sociolinguistic features as part of teaching

Table 3. Opening and Closing Moves in Chinese E-Letters

Greetings	Expressions	Count	Percent
Opening statement	Hello or Hi	16	73%
	Dear stranger	1	4%
	Nothing	5	23%
Closing formula	Hopefully	2	9%
	Sincerely or Yours sincerely	6	27%
	See you soon	1	5%
	Good luck to you	1	5%
	Nothing	12	54%

communicative competence. Belz (2005) goes further, with insights on how questioning styles or content can affect intercultural Internet interactions and with recommendations to language teachers to bring learners' awareness of their existing intercultural competence to the where these skills need to be taught. Table 3 shows ways in which Chinese students handled openings and closings in their letters of self-introduction: Ending a letter does not appear to be as smoothly done as opening one. More than half of the members of the group chose to send their letters without signaling a closure, contrary to the models given in most textbooks for courses relating to business communication and correspondence and certainly contrary to the models given collections of "canned" letters (Kaufer & Ishizaki 2006).

RESULTS: GAINING DESCRIPTIVE INSIGHTS

As mentioned above, we used the online corpus analysis tool, WMatrix, to identify the most frequent parts of speech in letters and midterms, dividing them by men and women, Chinese and American. Not surprisingly, self-reference is much more frequent in letters of self-introduction, regardless of gender or first language. It is reasonable to assume that both letters of self-introduction and definitions are informational, using Biber's distinction between informational and involved discourse (Biber, 1988). We hypothesize that, despite rapid changes in e-mail in the business world (Gimenez, 2006), more formal self-introductions would more likely be composed offline, would likely be characterized by the syntactic features of professional letters, and if sent electronically, might arrive as an attachment on e-mail stationery, whose background or letterhead has a special appearance. A good cross-cultural collaborative research project might be for writers from different areas of the world to review the use of e-mail stationery, for color, design, letterhead, and type font, to compare its use by gender, by business as opposed to personal venues, and the like.

Microanalysis of actual online usage in specific registers or genres makes a good companion to discussions such as that of Royal (2005), who summarizes journal-based scholarship on gender and the Internet. Politeness, or what we think is politeness, may be gender cued for a particular culture's use of a genre, as in the use of formulaic phrases in self-introductions. Our preliminary findings suggest the following considerations for teachers and editors working with pre-professional writing students or new hires, during training in formulaic sequence, idiom, or collocation derived from corpora, as suggested by Orr (2006).

First, if gender differences are associated with the creation or frequency of use of multiword expressions in a particular register or genre by native speakers/writers of a target language, such gender-based associations with specific areas of language production may be reflected in the first/dominant language of the nonnative speaker/writer, and such production (or its lack) may be misperceived or misappropriated as a genre or register cue by the second-language reader/speaker/writer. Instructors, editors, managers, and students have a great deal to contribute to the examination of gender differences by register or genre in workplace and professional settings, whether in the development, the implementation, or the analyses of corpora of professional texts by both novices and experts. Second, word-frequency lists are important linguistic tools coming out of corpus research, which can be used to great effect to improve the preparation of professional materials. For example, the most frequent formulaic expressions used by native speakers can be an important indicator for international editors, managers, and teachers in selecting or creating primary instructional materials for genre-based writing (on this, see Belcher, 2004; Silva & Brice, 2004).

DISCUSSION: SELF-REPRESENTATION, SOCIAL PROCESS, AND SMALL TALK ONLINE

Gender, genre, and style interact in formal writing, according to Argamon, Koppel, Fine, and Shimoni (2003), and in informal interaction, including text messages (Baron, 2004) and Web-noise according to reports of a study of mobile texting by Simeon Yates, Liberman (2005) compare this with studies of Weblog writings and the Gender Genie (2003) devised by Koppel and Argamon (as reported in Herring & Paolillo, 2006). Current research on Internet interaction among professionals on and off the job must address comments such as the following, by Wilson and Peterson (2002, p. 453): That is because technologies continue to evolve, even though general categories of communication will persist ". . . we are suggesting research that focuses on social processes and emerging communicative practices rather than on specific user technologies."

Accordingly, we turn to self-representation as social process and the need that we see for encouraging "small talk" prior to any kind of formal, Internet-supported interaction intent on joint tasks.

The work of researchers such as Eva W.-S. Lam (2004, 2006) reminds us that students in many countries at mid-decade—today's and tomorrow's technical writers—draw on a range of culturally contingent genres and styles to create their own interactional norms in what Bretag (2006), following Bhabha (1994), highlights as an online "third space." Third space occurs where a writer can reconstruct identity in interactions that otherwise might be unequal or asymmetrical. Bretag's (2006) discussion of praxis is a good case study of how this concept can be used for developing, maintaining, and sustaining interculturality in online interactions. Bretag studied a two-way e-mail exchange, finding, for example, that disclosure, defined as the sharing of intimate information, moved students toward a third space in which they could ask important questions and create their own norms for what seemed appropriate usage.

Even in something as formal as an MBA seminar, instructors will need to build in opportunities for personalization, socialization, and a real sense of "presence" if they are to achieve any kind of third space interculturality online. Warden, Chen, and Caskey (2005) report on cultural differences in online behavior between Chinese and Southeast Asian MBA students in an international MBA program at a university in Taiwan. The program also included Westerners. Asians and Westerners differed greatly in online behaviors, and Southeast Asian students differed from Chinese students, which the authors attributed to Chinese instructional traditions of a stronger master-student relationship, which grounds Chinese preferences for offering replies rather than initiating topics, and for ongoing efforts to protect face and smooth discord or controversy (Warden, Chen, & Caskey, 2005, p. 229). Perhaps most interesting was the authors' comment that both groups of Asian students differed from Westerners in being "happier to engage in small talk online, establishing context in the low-context medium," a comment that is followed by a warning against being reductive: "The common use of the English language may make an international class of students appear to be on the same wavelength, but . . . the meaning and use of teaching tools will be interpreted through each student's internalized value system" (p. 230).

Related to the above discussion are issues raised by Watson-Gegeo (2004). If we accept Watson-Gegeo's view that language is socially constructed and is shaped by cultural and sociopolitical processes, we need to rethink what happens in the classroom. Learners need to become analysts of language and therefore culture. They have to learn "multiple representations of culture." The reemergence of linguistic relativism, albeit in a new, expanded version, does offer one viable route for learners. The theory of linguistic relativism has wider implications for classroom organization and involves examining ways in which language is used to "create expectations, meanings, and judgments about learners, their knowledge(s) and indigenous/local/standpoint epistemologies . . . through interactional routines that invite while limiting agency" (Watson-Gegeo, 2004, p. 341).

Brown and Lewis (2003) offer a good illustration. Two teachers of a pre-employment English as a second language (ESL) course in New Zealand, Brown and Lewis reported on the recording and analysis of 10 hours of workplace conversation in a factory where many of the graduates from their course obtained employment. It is through research like this that our understanding of what language is and how it works in a variety of workplace settings across cultures can be continually refined and improved. One of Brown and Lewis's findings that attracted our attention was the fact that 50% of the discourse consisted of interpersonal social talk. The authors point out that a concentration on work-related topics in a traditional ESP course may leave the learners disadvantaged socially and affect their status. Small talk online can be initiated or closed off by initial self-representation and disclosure in the self-introduction. Presumably, a lack of the linguistic skills and cultural knowledge needed to engage in social talk would also affect the learners' success in obtaining the information they require, since the information holder's willingness to impart information and more importantly give extra assistance may in part be determined by her perceptions of the person seeking information formed as a result of previous social interaction or its lack.

This has an impact on the designing and understanding of synchronous and asynchronous Web-based interactions. Do we always build in time for social talk in e-mail exchanges or provide some sort of grounding or overview of the features of small talk? Perhaps what we need to do is to identify those linguistic and textual factors that lead to different stages of personal communication in e-mail exchanges. These may include some expression of commitment to the exchange that suggests a writer is willing to go beyond the initial stage of self-introduction. Can we identify a change in the level of personalization or of self-disclosure across cultures? How is this realized linguistically and pragmatically? For example, in letters of self-introduction, what may seem to members of one culture a sort of self-absorption, a focus almost entirely on selected aspects of a Self and preferred activities, may, to members of another culture, serve as an implicit invitation to self-disclose and thereby provide a means of establishing a basis for further communication.

When writers self-disclose in letters of introduction, they go beyond name, date, and serial number. The reader looks for some social sense. Figure 2 displays frequencies by percentages for selected semantic domains, as coded by WMatrix, across letters by Chinese and U.S. women. We looked for words signaling (self-)evaluation, emotion, social actions, and a subset of social actions: personality traits.

To get a sense of self-positioning in the self-introduction letters, we concordanced *I*-tokens, finding that the Chinese and American writers positioned themselves rather differently.

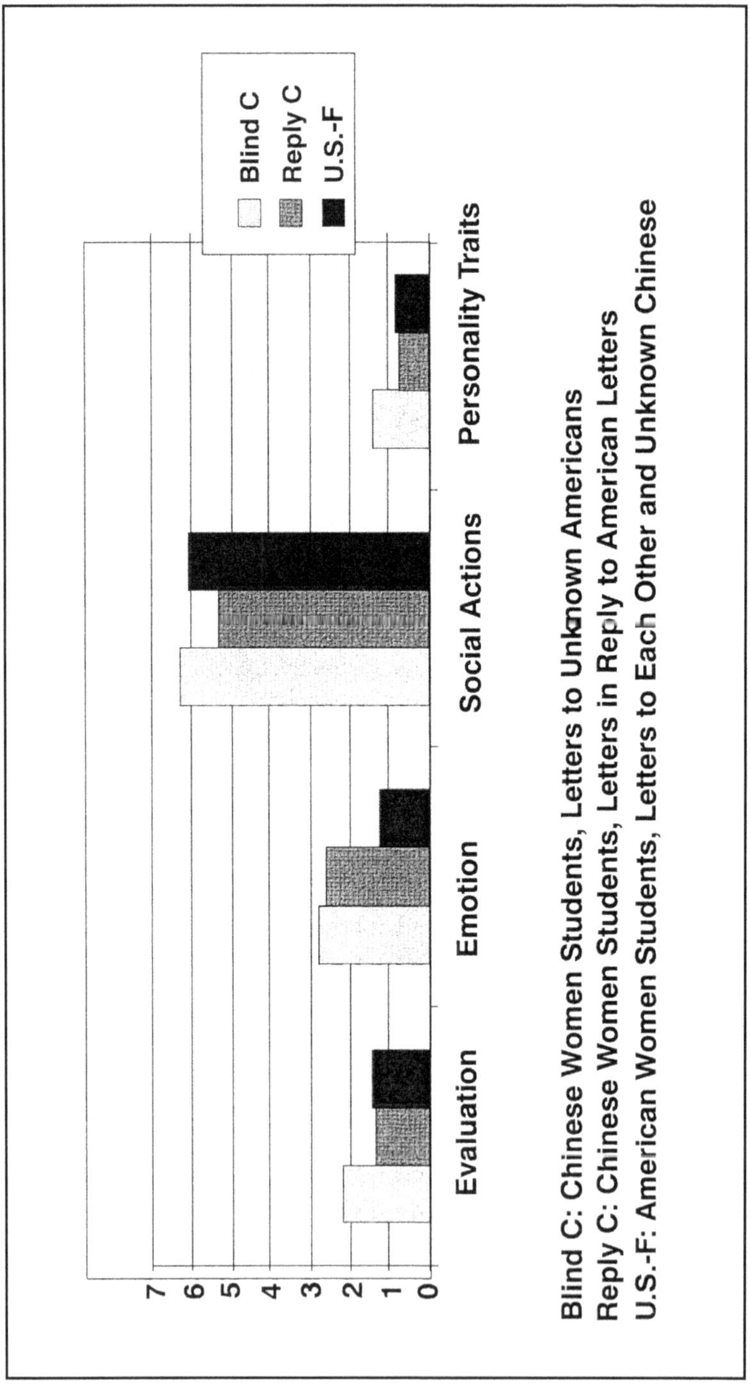

Figure 2. Semantic domains of evaluation, emotion, and social actions, states, and processes across three groups of female letter writers.

The American writers were taking an applied linguistics course that could be used toward a professional certificate in gerontology; their backgrounds included several different academic concentrations (nursing, gerontology, physiology and kinesics, communications, English as a second language, applied linguistics, and cognitivc sciencc). Thc Chincsc studcnts wcrc taking an advanced course in writing in English in the Applied Foreign Languages division of a college of international business; their backgrounds included education and business. Using the categories *autobiographical* identity and *discoursal* identity from Tang and John (1999) to arrange the *I*-phrases, we noted that American students used the discoursal *I* five times more often than the Chinese, who relied more heavily on the autobiographical *I*. Their autobiographical identity presented the life-story; their discoursal identity was keyed to current roles and relationships. The two groups also presented each other with different kinds of details.

- Americans presented themselves to each other and to the Chinese students with fewer personal details or self-disclosure, though their introductions were longer:

AUTOBIOGRAPHICAL WORK/PRE-UNIVERSITY	LOCATIONS AND COMPANIONS IN LIFE PRE-WORK/PRE-UNIVERSITY
Past Autobiography:	I'm __. I was born in/lived in __, went to school at __, grew up in/moved to __; my parents/grandparents live(d) in __.
Present Autobiography:	I have sisters/brothers/husband/roommates/ dog(s)/cat(s)/boy/girl friends; I live in a house/ dorm/apartment with ___.
DISCOURSAL	STUDENT, WORKER, LEARNER
Present-Day Studies:	I am taking/enrolled in/working on __ [course/degree].
Present Day Work:	I work at/work for/work part-time as __.
Future Work:	I plan to/hope to/___ after I graduate/finish an advanced degree.

- Chinese students, on the other hand, wrote shorter self-introductions and included different components in them:

AUTOBIOGRAPHICAL PREFERENCES	PERSONAL APPEARANCE, PREFERENCES
Past Autobiography:	I was born in ___; my parents/grandparents live(d) in ___; I have a large/small family.
Present Autobiography:	I look like ___; I like/enjoy music/sports/movies/ reading/travel.
DISCOURSAL	**PERSONAL CHARACTER TRAITS AS STUDENT, LEARNER**
Personal traits and welcome:	I'm a cheerful/optimistic/easy-going/happy/ friendly person; I look forward to knowing/ reading/writing/meeting/hearing from you;
Present-Day Studies:	I attend ___. I major in/study ___.
Present and Future Work:	I hope to teach/have a job in business; I teach English.

Because the preferences regarding self-presentation are so different in these sets of self-introductions, participants may have had little chance to identify common ground through the common categories surfacing in online small talk. Small talk is anything but little; as Coupland (2003, p. 1) comments, it "enacts social cohesiveness, reduces inherent threat values of social contact, and helps to structure social interaction." Sustained online interaction, such as collaboration in a writing task or a project, will benefit from some small talk among and across members. The groundwork for the social glue of small talk is laid down in the original e-exchange of self-introduction and the kinds of information that are disclosed in that exchange. The writing may be in a variety of English that all hold in common; however, the choices of what to disclose and how to disclose it are strongly culturally influenced.

In short, we find that the opportunity to provide scaffolding for small talk, to support some measure of self-disclosure, may be the crucial ingredient in successful online introductions and interactions. A brief discussion of the nature of small talk, perhaps accompanied by an exercise in brainstorming a short list of what could be included in an online self-introduction, can be useful to students as well as new professionals. They will benefit from knowing that their readers across the world will expect to learn some autobiographical details, including the size of their family, the area of the country that they live in, and the type of housing in which they live, and will also expect some discoursal details about their current major or work area, a job or a hobby they currently pursue, a book or movie they have enjoyed recently, and perhaps one small detail about their personality. As is reflected in the suggestions we provide below, we found the development of a small corpus of online interactions to be a fine resource for action research and materials development, as well as a way to better understand our students and their expectations of each other.

CONCLUSION: SOME IMPLICATIONS FOR FURTHER RESEARCH AND TEACHING

Our corpus-based insights led us to a greater concern about where and how to help writers in international settings edge out of their comfort zone, initiate mini-exchanges, and find common ground for public as well as private writing, from self-editing to commentary. We think the following are potentially valuable directions for research and teaching:

- Develop intercultural competence in online settings as a pedagogical priority. Korhonen (2004) describes an experiment in intercultural training with engineering students in Finland, using a package of "critical incidents" (scenarios) keyed to cultural issues. At the end of the experiment, "the students accept intercultural competence as an integral and equal part of the professional qualifications needed by Bachelors of Engineering in today's international and multicultural industry and business." This is paralleled in Liaw's (2007) work with engineering students in Taiwan.
- Use corpus-enriched work to look not only at specific features of language but also at their function, particularly in the formation of online international communities of practice.
- Examine the context in which particular techniques or technologies will not work in a particular culture, and the reasons for this. Fang and Warschauer (2004); Kern, Ware, and Warschauer (2004); and Dautermann (2005) all provide fine examples of ways to learn from disappointments, as they review a number of challenges, failures, and similar opportunities to grow.
- Identify and explore key features of discourse as situated within discourse communities. A striking example of linguistic/rhetorical differences between current Anglo-American writing and Chinese writing is the use of the enumerated list, a common feature of Chinese writing. Writing textbooks give little advice on when and how to use lists, and most model essays do not contain lists. Such lists, however, may be found in the *gong wen* (official, public documents) used in most administrative organizations, which may have their origin in earlier forms of Chinese writing.
- Examine commonalities across genres and registers. For example, Kong (2005, p. 290) offers a fine-grained study of ways in which Chinese and English writers present evaluation in writing, using similar grammatical structures but to different degrees.

Discussions of English as a lingua franca for international meetings, such as that of Rogerson-Revell (2006), emphasize the "functional realism" of current research, and the growth of discussions about cultural training in the literature on international management (Rogerson-Revell, 2006, p. 6). In Nickerson's review of English as the lingua franca of international business, she notes the emphasis on "one of four major communicative genres: negotiations, meetings, e-mail and business letters" (2005, p. 369). Indeed, a large amount of Internet-supported, cross-cultural professional writing is done in English. In learning how to write professional and technical documents in English, and in working with them in online settings, are students facing a language problem or a writing problem? The answer, of course, is both, in that writing in English presents the writer with the problem of identifying the discourse type, register, or genre that is needed, and its appropriate linguistic and rhetorical features.

What remains less clear, as English continues to develop as a global lingua franca, is the extent to which Anglo-American English will remain the dominant language variety, particularly on the Internet (see Danet & Herring 2007), or the language type that learners aspire to, as other cultures challenge the Anglo-American construction of language and knowledge (Watson-Gegeo, 2004). How will the exploding numbers of online netizens affect the way in which online discourse communities are constructed and maintained? Surely, a great deal of this participation will take place among professionals in the blogosphere, on the Web site, through the podcast, in the e-mail box, throughout cyberspace, across the ether, virtually next door. Instructors of technical writing may not choose to develop corpora of online interactions as part of their pedagogical repertoire, but they and their students will benefit from corpus-based analyses and data arrays that index subtle cultural constructs such as self-presentation, politeness phrases, and when or how to offer and respond to different question types. Online writing has social features, but most researchers agree that it is not socially transparent, leaving more room than we like for cultural misunderstandings. Learning how to meet one other online is a useful, perhaps essential, skill.

REFERENCES

Al-Khatib, M. A. (2001). The pragmatics of letter-writing. *World English 20*, 179–200.

Angelova, M., & Riazantseva, A. (1999). "If you don't tell me, how can I know?" A case study of four international students learning to write the U.S. way. *Written Communication, 16,* 491–525.

Argamon, S., Koppel, M., Fine, J., & Shimoni, A. (2003). Gender, genre, and writing style in formal written texts. *Text, 23,* 321–346.

Artemeva, N. (1998). The writing consultant as cultural interpreter: Bridging cultural perspectives on the genre of the periodic engineering report. *Technical Communication Quarterly, 7,* 285–299.

Baron, N. (2004). See you online: Gender issues in college student use of instant messaging. *Journal of Language and Social Psychology, 23,* 397–423.

Belcher, D. (2004). Trends in teaching English for specific purposes. *Annual Review of Applied Linguistics, 24,* 165-186.

Belz, J. A. (2005). Intercultural questioning, discovery and tension in Internet-mediated language learning partnerships. *Language and Intercultural Communication, 5,* 3–37.

Bhabha, H. K. (1994). *Location of culture.* London: Routledge.

Biber, D. (1988). *Variation across speech and writing.* Cambridge, MA: Cambridge University Press.

Biber, D. (1994). Using register-diversified corpora for general language studies. *Computational Linguistics, 19,* 219–241.

Bretag, T. (2006). Developing "third space" interculturality using computer-mediated communication. *Journal of Computer-Mediated Communication, 11*(4). Retrieved June 2007, from http://jcmc.indiana.edu/vol11/issue4/bretag.html

Brown, T., & Lewis, M. (2003). An ESP project: Analysis of an authentic workplace conversation. *English for Specific Purposes, 22,* 93–98.

Campbell, C. (1998). Rhetorical ethos: A bridge between high-context and low-context culture. In S. Niemeier, C. Campbell, & R. Dirven (Eds.), *The cultural context in business communication* (pp. 31–47). Philadelphia: John Benjamins.

Connor, U. (1996). *Contrastive rhetoric: Cross-cultural aspects of second-language writing.* Cambridge, MA: Cambridge University Press.

Corpus of Professional English. (in progress). Retrieved June 2007, from www.perc21.org/cpe-project

Coupland, J. (2003). Small talk: Social functions. *Research on Language and Social Interaction, 36,* 1–6.

Coverdale-Jones, T. (2006). Afterword: The Chinese learner in perspective. *Language, Culture and Curriculum, 19,* 148–153.

Crystal, D. (2001). *Language and the Internet.* Cambridge, MA: Cambridge University Press.

Danet, B., & Herring, S. (Eds.). (2007). *The multilingual Internet: Language, culture, and communication online.* New York: Oxford University Press.

Daoud, S. (1998). How to motivate EFL learning and teaching of academic writing by cross-cultural exchanges. *English for Specific Purposes, 17,* 391–412.

Dautermann, J. (2005). Teaching business and technical writing in China. *Technical Communication Quarterly, 14,* 141–159.

Davis, B., & Brewer, J. (1997). *Electronic discourse: Linguistic individuals in virtual space.* Albany, NY: State University of New York Press.

Davis, B., & Mason, P. (2006). Trying on voices: Using questions to establish authority, identity, and recipient design in electronic discourse. In R. Scollon & P. LeVine (Eds.), *Discourse and technology: Multimodal discourse analysis. Georgetown University Round Table in Linguistics, 29* (pp. 34–46). Washington, DC: Georgetown University Press.

Davis, B., & Thiede, R. (2000, 2006). Writing into change: Style-shifting in asynchronous electronic discourse. In M. Warschaeur & R. Kern (Eds.), *Network-based teaching: Concepts and practice* (pp. 87–120). Cambridge, MA: Cambridge University Press. (Original work published 2000).

ENGICORP. (in progress). Illinois Institute of Technology. Retrieved June 2007, from www.iit.edu/~tcid/engicorp

Fang, X., & Warschauer, M. (2004). Technology and curricular reform in China: A case study. *TESOL Quarterly, 38*(2), 301–323.

Garside, R., & Smith, N. (1997). A hybrid grammatical tagger: CLAWS4. In R. Garside, G. Leech, & A. McEnery (Eds.), Corpus annotation: Linguistic information from computer text corpora (pp. 102–121). London: Longman.

Gender Genie. (2003). Accessed June 1, 2007, from http://www.bookblog.net/gender/genie.html

Giammona, B. (2004). The future of technical communication: How innovation, technology, information management, and other forces are shaping the future of the profession. *Technical Communication, 51*, 349–366.

Gimenez, J. (2006). Embedded business e-mails: Meeting new demands in international business communication. *English for Specific Purposes, 25,* 154–172.

Goby, V. (2007). Business communication needs: A multicultural perspective. *Business and Technical Communication, 21,* 425–439.

Graddol, D. (2006). *English next.* Retrieved from http://www.britishcouncil.org/learning-research-english-next.pdf

Greenfield, R. (2003). Collaborative e-mail exchange for teaching secondary ESL: A case study in Hong Kong [Electronic version]. *Language Learning and Technology, 7,* 46–70.

Herring, S., & Paolillo, J. (2006). Gender and genre variation in Weblogs. *Journal of Sociolinguistics, 10,* 439–459.

Holland, R. (2002). Globospeak? Questioning text on the role of English as a global language. *Language and Intercultural Communication, 2*, 5–24.

Holmes, J. (2006). *Gendered talk at work: Constructing gender identity through workplace discourse.* Oxford, England: Blackwell.

Hondo, J., & Goodman, B. (2001). Cross-cultural varieties of politeness. *Texas Papers in Foreign Language Education, 6*, 163–170.

Kankaanranta, A. (2006). "Hej Seppo, could you pls comment on this!": Internal e-mail communication in lingua franca English in a multinational company. *Business Communication Quarterly, 69,* 216–225.

Kaufer, D., & Ishizaki, S. (2006). A corpus study of canned letters: Mining the latent rhetorical proficiencies marketed to writers-in-a-hurry and non-writers. *IEEE Transactions on Professional Communication, 49,* 254–266.

Kern, R., Ware, P., & Warschauer, M. (2004). Crossing frontiers: New directions in online pedagogy and research. *Annual Review of Applied Linguistics, 24,* 243–260.

Kessapidu, S. (1997). A critical linguistic approach to a corpus of business letters in Greek. *Discourse and Society, 8,* 479–500.

Kirkpatrick, A. (1991). Information sequencing in Mandarin in letters of request. *Anthropological Linguistics, 33,* 1–20.

Kong, K. (2005). Evaluative resources in English and Chinese research articles. *Multilingua, 24,* 275–308.

Korhonen, K. (2004). Developing intercultural competence as part of professional qualifications: A training experiment. *Journal of Intercultural Communication, 7.* Retrieved June 2007, from http://www.immi.se/intercultural/nr7/korhonen-nr7.htm

Kumaravadivelu, B. (2003). Problematizing cultural stereotypes in TESOL. *TESOL Quarterly, 37,* 709–719.

Lam, E. (2004). Second language socialization in a bilingual chat room: Global and local considerations. *Language Learning and Technology, 8*(3), 44–65.
Lam, E. (2006). Culture and learning in the context of globalization: Research directions. *Review of Research in Education, 30*. Retrieved October 2006, from http://www.sesp.northwestern.edu/docs/publications/202476101446324778 6e208.pdf
Lavid, J., & Taboada, M. (2004). Stylistic differences in multilingual administrative forms. *Journal of Technical Writing and Communication, 34*, 43–65.
Lee, D., & Swales, J. (2006). A corpus-based EAP course for NNS doctoral students. *English for Specific Purposes, 25*, 26–75.
Leech, G. (1983). *Principles of pragmatics*. New York: Longman.
Li, Y., & Flowerdew, J. (2007). Shaping Chinese scientists' manuscripts for publication. *Journal of Second Language Writing, 16*, 100–117.
Liaw, M. (2007). Constructing a "third space" for EFL learners: Where language and culture meet. *ReCALL, 29*, 224–241.
Liberman, M. (2005). "Twice as long" is 50% longer, or may be 42% longer, or was it 21% longer? *Language Log*, November 15, 2005. Retrieved February 1, 2009, from http://itre.cis.upenn.edu/myL/languagelog/archives/002649.html
Ma, R. (1996). Computer-mediated communication between East Asian and North American college students. In S. Herring (Ed.), *Computer-mediated communication: Linguistic, social and cross-cultural perspectives* (pp. 173-185). Philadelphia: John Benjamins.
Matsuda, P., Canagarajah, A., Harklau, L., Hyland, K., & Warschauer, M. (2003). Changing currents in second language writing research. *Journal of Second Language Writing, 12*, 159–179.
Mondorf, B. (2002). Gender differences in English syntax. *Journal of English Linguistics, 30*, 158–180.
Nickerson, C. (2005). English as a lingua franca in international business contexts. *English for Specific Purposes, 24*, 367–380.
Orr, T. (2006). Introduction to the special issue: Insights from corpus linguistics for professional communication. *IEEE Transactions on Professional Communication, 49*, 213–216.
Ortiz, L. (2005). Cruzando las fronteras de la comunicación profesional entre México y los Estados Unidos: The emerging hybrid discourse of business communication in a Mexican-U.S. border region. *Journal of Business Communication, 42*, 28–50.
Rayson, P. (2007). *Wmatrix: A web-based corpus processing environment*. Lancaster, England: University of Lancaster. Retrieved June 2007, from http://www.comp.lancs.ac.uk/ucrel/wmatrix/
Rayson, P., Archer, D., Piao, S. L., & McEnery, T. (2004). The UCREL semantic analysis system. In the proceedings of the LREC 2004 Workshop, *Beyond Named Entity: Recognition Semantic Labelling for NLP Tasks* (pp. 7–12). Lisbon, Portugal. Retrieved June 2007, from http://eprints.lancs.ac.uk/12453/1/usas_lrec04ws.pdf
Rogerson-Revell, P. (2006). Research note: Using English for international business: A European case study. *English for Specific Purposes, 26*, 103–120.
Royal, C. (2005). A meta-analysis of journal articles intersecting issues of Internet and gender. *Journal of Technical Writing and Communication, 35*, 403–429.
Scollon, R., & Scollon, S. (2000). *Intercultural communication* (2nd ed.). Oxford, England: Blackwell.

Silva, T., & Brice, C. (2004). Research in teaching writing. *Annual Review of Applied Linguistics, 24,* 70–106.

Sinclair, J. (2004). *How to use corpora in language teaching.* Amsterdam: John Benjamins.

Someya, Y. (2000). Yasumasa Someya's learner business letters corpus. Retrieved June 2007, from http://ysomeya.hp.infoseek.co.jp

St.Amant, K. (2002). When cultures and computers collide: Rethinking computer-mediated communication according to international and intercultural communication expectations. *Journal of Business and Technical Communications, 16,* 196–214.

Tang, R., & John, S. (1999). The "I" in identity: Exploring writer identity in student academic writing. *English for Specific Purposes, 18,* 23–39.

Upton, T. A., & Connor, U. (2001). Using computerized analysis to investigate the text linguistic discourse moves of a genre. *English for Specific Purposes, 20,* 313–329.

Warden, C., Chen, J., & Caskey, D. (2005). Cultural values and communication online: Chinese and Southeast Asian students in a Taiwan International MBA class. *Business Communication Quarterly, 68,* 222–232.

Warschauer, M. (1999). *Electronic literacies: Language, culture, and power in online education.* Mahwah, NJ: Erlbaum.

Warschauer, M., & Kern, R. (Eds.). (2000). *Network-based language teaching: Concepts and practices.* New York: Cambridge University Press. (Reissued 2006).

Warschauer, M., Shetzer, H., & Meloni, C. (2000). *Internet for English teaching.* Philadelphia: TESOL Publications.

Watson-Gegeo, K. A. (2004). Mind, language, and epistemology: Toward a language socialization paradigm for SLA. *Modern Language Journal, 88,* 331–350.

Wilson, S., & Peterson, K. (2002). The anthropology of online communities. *Annual Review of Anthropology, 31,* 449–671.

Wolverhampton Business English Corpus. Retrieved June 2007, from www.elda.org/catalogue/en/text/W0028.html

Wong, A. (2005). Writer's mental representations of the intended audience and of the rhetorical purpose for writing and the strategies that they employed when they composed. *System, 33,* 29-47.

Woolever, K. (2001). Doing global business in the information age: Rhetorical contrasts in the business and technical professions. In C. Panetta (Ed.), *Contrastive rhetoric revisited and redefined* (pp. 47–64). Mahwah, NJ: Erlbaum.

Xiao, H., & Petraki, E. (2007). An investigation of Chinese students' difficulties in intercultural communication and its role in ELT. *Journal of Intercultural Communication, 13.* Retrieved June 2007, from http://www.immi.se/intercultural/

Zhu, W. (2000). Rhetorical moves in Chinese sales genres, 1949 to the present. *Journal of Business Communication, 27,* 156–172.

Zhu, W. (2004). Writing in business courses: An analysis of assignment types, their characteristics, and required skills. *English for Specific Purposes, 23,* 111–135.

SECTION III

Cross-Cultural Collaborations and Learning Environments

CHAPTER 8

Virtual Design Studio: Facilitating Online Learning and Communication Between U.S. and Kenyan Participants

Audrey Bennett, Ron Eglash, and Mukkai Krishnamoorthy

CHAPTER OVERVIEW

This chapter analyzes the way in which existing instructional and communication technologies can bridge geographic, digital, and cultural divides and facilitate collaboration on an international scale. It highlights some existing Web-based technologies for distance learning and synchronous and asynchronous communication; and it explores their limitations in terms of collaboration between first- and third-world participants. Three disciplinary perspectives—those of visual communication, anthropology, and computer science—are presented at times in dialogue with each other. The synthesis of these three perspectives offers technical communicators an interdisciplinary understanding of important intercultural and technical issues involved in international online learning and health-related communication.

In the summer of 2003, three multidisciplinary educators and one graduate student conducted a collaborative design workshop with laypeople in Kenya in a rural community resource center. In a participatory manner, a local group of bilingual (English/Luo) Kenyans designed HIV/AIDS prevention and awareness posters for their community. Simultaneously, by way of a *virtual design studio*

(VDS) constructed out of existing communication technologies (e.g., e-mail, online chat rooms, collaborative-learning software), the educators situated in front of their computer screens in Troy, New York, indirectly observed and participated in the Kenyans' design process. Understanding the intercultural communication and technical issues of international collaborations of this kind is the next frontier for the technical communication discipline as it broadens to embrace the opportunities and challenges that globalization offers. In "The Future of Technical Communication" (2004), Giammona argues that the skill set of the future technical communicator will include the ability to learn to use new tools and work closely with information developers in different languages and from different cultural perspectives. This chapter aims to disclose and address these important issues through an analysis of the aforementioned virtual design studio as a shared context for cross-cultural communication and collaboration between the first and third worlds. It shows how third-world laypeople and first-world educators collaborated and designed an HIV/AIDS awareness campaign through the use of existing communication technologies for distance learning and synchronous and asynchronous communication.

AIDS IN AFRICA AS A CROSS-CULTURAL, MULTIDISCIPLINARY PROBLEM

HIV/AIDS is a humanitarian crisis and global pandemic that threatens our political and economic stability. In the absence of a vaccine (and probably even in its presence), research shows that communication is key to HIV/AIDS prevention (Choi & Coates, 1994; Kiwanuka-Tondo & Synder, 2002). According to Forslund (1996) and Andrews (2000), print communication is particularly effective with nonreading international audiences—like third-world laypeople—who rely heavily on image-based information to retrieve and remember important messages. The design of HIV/AIDS prevention campaigns in the United States, for instance, has benefited from the participation of multidisciplinary experts including technical communicators, graphic designers, advertising agencies, and others. However, although there have been successful examples of HIV/AIDS prevention interventions in Africa, like Kiwanuka-Tondo and Synder's efforts in Uganda (2002), African rural regions, which have the highest rate of new infections and overall number infected, are largely lacking in such expertise and production resources. Experts in the United States who want to assist with communicating HIV/AIDS awareness and prevention information to an African audience face a cross-cultural, transnational communication gap between themselves and African laypeople. That is, the U.S. experts lack the cultural understanding of African societies (language, colloquialisms, visual rhetoric, etc.) that would allow them to effectively communicate HIV/AIDS prevention messages to African audiences. Indeed, previous studies of HIV/AIDS prevention print campaigns in Africa have shown that the successful communication conventions are those in which the visual designs were influenced by members of their local target audience (Kiwanuka-Tondo & Synder, 2002). The primary reason, Forslund confirms in "Analyzing Pictorial Messages Across Cultures" (1996), is that culture influences greatly the way in which nonreaders (including people in third-world contexts) perceive information. Thus, instead of tailoring images to the culture of the target audience (Hager, 2000), we posit that first-world experts can ameliorate cross-cultural communication problems like culturally ambiguous imagery by bringing third-world laypeople into the design process and allowing them to choose or create their own images.

Since we are not located in Africa, we decided to approach this design collaboration through instructional and communication technologies for distance learning and communication; a system that has been referred to by Wojtowicz (1995) and by Scrivener, Ball, and Woodcock (2000) as a virtual design studio (VDS). These authors' idea of a VDS is that it facilitates collaborative designing between students and faculty located in different parts of the world. A VDS system, however, has the potential to facilitate unorthodox educational training by allowing first-world educators in locations far from the third world (in our case, in the United States) to interact directly with third-world laypeople in Africa. With the dramatic increase in African Internet access, it has now become technically feasible to provide a system of virtual collaboration and educational training that will enable and empower African laypeople to develop their own AIDS awareness and prevention communication forms. Figure 1 illustrates some of the existing communication forms found in Kenya.

MULTIDISCIPLINARY PERSPECTIVES

The team comprised a graphic design researcher (Audrey Bennett), an anthropologist in science and technology studies (Ron Eglash), a computer scientist (Mukkai Krishnamoorthy), and a doctoral student in science and technology studies (Marie Rarieya), and thus three disciplinary perspectives were represented in this research project: visual communication, anthropology, and computer science. There were both differences and similarities between the objectives of each educator. The next four sections will present their different perspectives in relation to findings and perspectives in technical communication.

Toward a Framework for Virtual Collaborative Learning with International Participants: A Multidisciplinary Perspective

In recent history, various disciplines including technical communication have begun to recognize the importance of online learning and collaboration with international participants (Cherkasky, Greenbaum, Mambrey, & Pors, 2000; Giammona, 2004; Herrington & Tretyakov, 2005; Scrivener et al., 2000; Shuler & Namioka, 1993). Our desire to include third-world laypeople in a user-centered

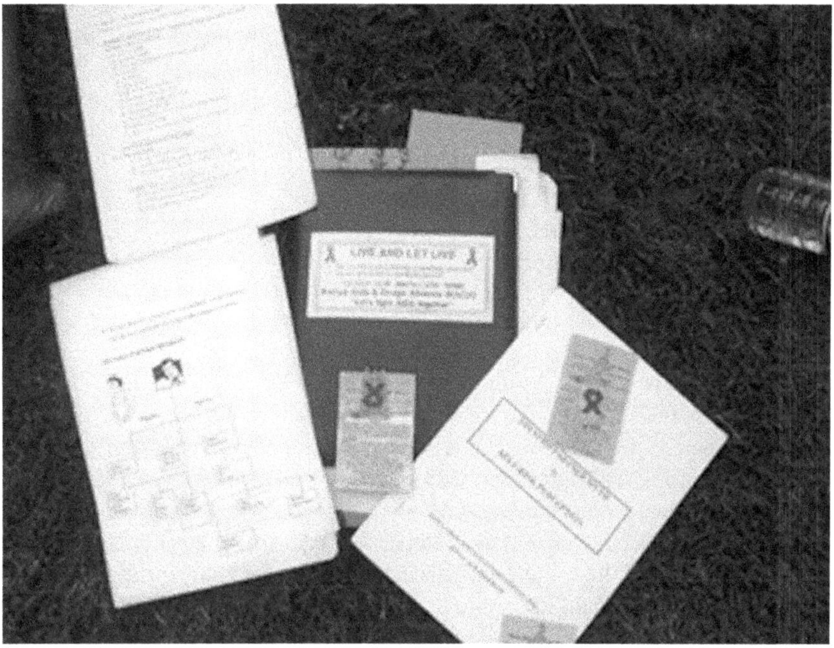

Figure 1. Examples from "Live and Let Live," an AIDS/HIV awareness and prevention campaign found in Kenya prior to our collaborative workshop. The chart on the left shows an infection chain.

pedagogical graphic design project presented various logistical, technical, and cultural issues that led us to examine documented user research models for design that have been proven effective through practical application. These are typically qualitative and quantitative methods that include questionnaires, surveys, and focus groups. We found that though most of these methods often yield indispensable information, they tend to keep the designer (instead of the audience) in control of the concept and the final form—a potential problem due to the issues, noted above, related to communicating verbal and visual information across cultures.

The discipline of technical communication has broken ground in the area of international collaborative education facilitated by online technology. For instance, in the "Global Classroom Project" (GCP), Herrington and Tretyakov (2005) use Web-based conferencing software to facilitate multimodal discussion sessions around common readings and analytical projects. The use of a Web-based communication software program called WebBoard allows participants to post visual materials as well as hypertext. Based on an egalitarian theme, GCP gives students control over their writing projects in the course (Herrington & Tretyakov, 2005). Like the GCP, our project gives students—that is, Kenyan lay participants—control over the final outcome of the design process. However, our project is novel in that it uses both an on-site participatory design workshop and a VDS that permits image editing and creation.

Toward a Cross-Cultural, Visual Aesthetic: A Graphic Design Perspective

Research in graphic design and technical communication shows that images, colors, typefaces, symbols, and organizational formats communicate different things to different audiences (Coe, 2001; Forslund, 1996; Frascara, 1997; Horton, 2005; Lipton, 2002). A cross-cultural graphic created for informational purposes is communicatively effective when its visual aesthetics resonate with the target audience's culture. The questions that arose for us, then, were the following: What is the visual vernacular of Kenyans? Are there any intercultural visual symbols or codes that are shared by both Kenyan and American audiences? Thus, our objective was to determine the visual vernacular of Kenyan laypeople and find out whether we could use some universal visual symbols and codes or only culturally specific ones.

Indigenous Knowledge Systems: An Anthropological Perspective

Previous fieldwork in Africa has demonstrated valuable educational applications for computational models of indigenous knowledge (Eglash, 1999). These educational materials have been of particular importance for U.S. minority students in addressing their lack of engagement and low performance in math and science curricula. Students have shown a tendency to see their own cultural identity as antagonistic to the curriculum. The barriers include myths of genetic determinism and conflicts with stereotypes of cultural authenticity or social relevance (Downey & Lucena, 1997; Geary, 1994; Ogbu, 1998). By offering minority students a computational medium ("Culturally Situated Design Tools") in which indigenous and vernacular designs can be created using mathematics, we have found possibilities for a less alienating experience (Eglash, Bennett, O'Donnell, Jennings, & Cintorino, 2006). Similarly, the VDS representations of local Kenyan concepts through print and electronic media could result in reducing alienation from HIV/AIDS educational materials—a hypothesis that aligns with Forslund's (1996) and Andrews' (2000) theories on the effectiveness of print communication. In ways that bear some relation to the U.S. minority experience, oppressive histories involving the colonial past and neocolonial present have created barriers for African populations, as in the suspicion with which condom use is often regarded.

Systemic Models of HIV/AIDS in Kenyan Society: A Computer Science Perspective

Computational theory from computer science has been used to model viral spread in human populations, including the spread of AIDS (Barabasi, 2002). How can these sophisticated models interact with our knowledge-based efforts? How might such network models be translated into indigenous understandings of kinship networks or other traditional concepts? Is it possible that playing simulation games in which one can visualize a network of viral infection could substantially change an individual's view of HIV risk?

While technical approaches tend to focus on the virus, increasingly researchers have pointed out that prevention depends on understanding the complex cultural, political, and economic dimensions of infection. For example, a recent article in *Physical Review* by Dezso and Barabasi (2002) suggested that a network model showed a drastic reduction of HIV rates when highly promiscuous individuals were removed from the network.

Highly connected nodes are a source of vulnerability in scaling networks. While this implies problems for networks we would like to keep—for example, problems like the vulnerability of the Internet to terrorist attacks—it implies solutions for networks we don't like, such as the network of AIDS infection. But translating this scaling network effect directly into social policy is problematic. In Africa, the "witch hunt" for infected individuals often leads to violent acts, which then create an atmosphere of fear. Those who might otherwise seek testing and prevention are cowed into inaction and silence. The result is an increased infection rate. On the other hand, the role of computer science in providing a computational design medium seems beneficial here; it allows a flexibility in which local design sensibilities can be expressed. Free expression in the context of HIV/AIDS education would help to solve the problem of inaction and silencing.

APPROPRIATING THE VIRTUAL DESIGN STUDIO MODEL FOR CROSS-CULTURAL AND -DISCIPLINARY COLLABORATION

The term virtual design studio has been used by many researchers in architecture, graphic design, and other fields to describe how computer media can support a collaborative process (cf. Cherkasky et al., 2000; Scrivener et al., 2000; Wojtowicz, 1995). While the innovations have typically focused on collaboration within educational institutions, the VDS could potentially facilitate the participation of remote laypeople and grassroots organizations. Collaboration in design, via a VDS, with grassroots organizations can vary across a broad spectrum of practices, and in its weakest form is merely glorified market research. But the potential for collaborative learning to truly empower marginalized communities worldwide—to facilitate their ability to gain greater independence and quality of life—has been demonstrated in a wide variety of development projects. In our project, we aimed to develop a VDS out of existing technologies that would facilitate the synchronous and asynchronous design of an HIV/AIDS prevention campaign by collaboration between us and third-world laypeople by way of partnering with nongovernmental organizations (NGOs) in rural Kenya.

Envisioning the Technological Infrastructure of the VDS

The VDS model that we envisioned at the beginning of our research allows a team of educators situated geographically in the United States to carry out synchronously and, at times, asynchronously the systematic design of an HIV/AIDS campaign for and with a target audience in Kenya. Central to the communication design process—particularly with reference to topics that bring up both psychologically and sociologically volatile issues such as AIDS—is face-to-face interaction. In design studio classrooms, for instance, instructors are often gauging the emotional state of students through both verbal and nonverbal cues as the students describe their work, and must take these affective investments into account in guiding the students' next steps—which the instructor can observe by looking over the students' shoulders as they work. Our vision of the VDS includes digital communication media that sustain this important aspect of instructor/student interaction, in addition to the technical demand that all participants brainstorm, conceptualize, sketch, render, and critique prototypes collaboratively. At a minimum, a VDS should enable U.S. educators and laypeople in Kenya to view a sketch simultaneously and make comments via a live chat space on the same screen juxtaposed to the sketch, and it should also allow the integration of drawing tools (like those found in industry-standard design software applications such as Adobe Photoshop or Illustrator) that enable the participants to modify visual parameters or edit the sketch in real time. Even this minimal description is different from previous attempts to set up a virtual design studio in an online learning situation, in which professional-to-professional or professional-to-student communication encourages downloading a *virtual whiteboard,* modifying it on a hard drive, and uploading the modified image. The real-time interaction demands of design instruction are not typically taken into account in such projects. Even those few VDS systems that attempt real-time collaborative editing—found more in the computer-supported cooperative work (CSCW) literature than in the communication design and visual communication literature—do not consider a lay audience or a third-world context. Our system is innovative in its emphasis on synchronous collaborative visual design with laypeople in the third world. If we use communication technologies to construct a virtual design studio that facilitates collaborative designing with members of the target audience, we can forge a virtual bridge between marginalized communities and ourselves across the globe.

With a Rensselaer seed grant in hand, we set out to technically facilitate a part virtual, part face-to-face collaborative design workshop in Kenya. We explored existing Internet-based environments like Yahoo Groups and collaborative learning software programs (e.g., Learnlinc) that were close in functionality to what we envisioned for our virtual design studio.

TECHNICAL FACILITATION OF COLLABORATION BETWEEN KENYAN LAYPEOPLE AND U.S. EDUCATORS ON THE DESIGN OF COMMUNICATION MATERIALS

In spring 2003, prior to the workshop in Kenya, ethnography and contextual design (Beyer & Holtzblatt, 1998) of face-to-face instruction in Bennett's design studio course at Rensselaer Polytechnic Institute (RPI) were used to sketch out a framework for some of the technical specifications of the VDS we envisioned. The first formative evaluation used student volunteers, an instructor, and external observers to compare the face-to-face design process in the physical classroom with the same tasks accomplished by existing collaborative communication systems. Specifically, we evaluated an existing Internet-based communication system (accessible gratis) called Yahoo Groups.

Yahoo Groups (see www.groups.yahoo.com/group/rvds)

Students in Bennett's studio design course participated in preliminary testing within a Yahoo group in which they communicated with Bennett, Eglash, and Krishnamoorthy on the creative development of their concepts using Yahoo's synchronous chat component. Using the posting, file sharing, and chat facilities provided by the Yahoo newsgroup, the communications were both observed in real time and recorded (by way of screen shots) for later analysis. In this study, we analyzed the impact of discussing purely in ASCII text, in contrast to the classroom experience of the same small group gathered around a computer screen. Initial analysis indicates that the lack of images in the newsgroup, the inability to modify an image under discussion, and the lack of nonverbal cues are all detrimental to this particular communication design activity. We also determined that while Yahoo allows for synchronous communication, the computer-mediated discussions progressed at a much slower pace than classroom discussions. While group communication among three or more people was encouraged through Yahoo, it was not well facilitated by the chat component.

In summer 2003, the actual collaborative workshop was held, using a combination of local instruction in the Kenyan community computing center (see Figure 2) and distance education resources at Rensselaer. We literally attempted to bring two different parts of the world together in a virtual space. We adapted, on a small scale, the communication design studio pedagogy to the Kenyan context. This workshop, which served as the second formative evaluation of existing technologies for the VDS, was approved by RPI's internal review

Figure 2. Community resource center in Kenya where the on-site and VDS workshop took place.

board and funded by a Rensselaer seed grant that enabled us to test collaborative learning software in a participatory design pilot study in Kenya. We held our small-scale design workshop in Kenya in collaboration with our partners, the International Center for Research in Agroforestry (ICRAF) and the Regional Land Management Unit (RELMA). The latter is an NGO in Nairobi, Kenya, that is funded by the Swedish International Development Cooperation Agency (SIDA).

Rarieya, who had previously taken Bennett's communication design studio course, set up a similar studio course structure for the design workshop but with only five participants (as opposed to Bennett's course, which enrolled over 25 students). Using the same course theme, communication design for AIDS prevention, Rarieya adapted Bennett's course to a very different context. Bennett, Eglash, and Krishnamoorthy at RPI attempted to interact in real time with the Kenyan participants through various instructional and communication technologies.

Upon her arrival in Kenya, Rarieya recruited participants through ICRAF. She created two sites for the workshop: Nairobi and Kusa. The latter was facilitated by RELMA. After she confirmed the sites, recruited participants, and started the workshop, we tested the feasibility of a set of existing communication technologies in terms of their capacity to facilitate collaborative learning in visual communication with our remote grassroots audience at the two sites in Kenya. Using a combination of low-tech media (cameras, sketchpads) on-site in Kenya and high-tech, Internet-based technologies (E-mail, Web sites, chat rooms, electronic whiteboards, industry-standard software applications for page layout and image-editing, etc.) for collaboration between the Kenyan sites and our interdisciplinary team of educators at RPI, we collaboratively designed an HIV/AIDS awareness and prevention campaign that included posters and post-cards for use in print and potentially in electronic media.

E-Tool for Collaborative Learning: Learnlinc

Learnlinc uses synchronous and asynchronous communication and offers interactive components that allow design students and their instructor to engage in group critiques in real time where they see, discuss, and modify sketches and prototypes as they would in the real classroom setting. Some of the relevant interactive components included in this software are the following:

- 2-way, Voice-Over Internet Protocol (VoIP)
- shared interactive whiteboard
- instant messaging
- application sharing
- breakout rooms
- interactive quiz and survey manager
- PowerPoint import
- on-the-fly content creating and editing tools
- Web push
- record and playback
- instructor control
- instant polling
- classroom-pace feedback

Though Learnlinc states that it allows application sharing, we were unable to open Adobe Photoshop—a key image-editing software program for graphics—during this formative evaluation session. We were able to connect minimally with Rarieya and the Kenyan participants through Learnlinc during the summer workshop. With an Internet connection speed of only 56K in Kenya, we were able to open Learnlinc and chat through instant messaging about each participant's design work. However, we were unable to use the graphical components due to the low bandwidth connection in Kenya.

E-Tools, Yahoo Group

The U.S. team also communicated with Rarieya through e-mail, cell phones (as needed and feasible), and a Yahoo group called "Rensselaer Virtual Design Studio (RVDS)" weekly, though we were not able to communicate directly with the participants in the Yahoo group primarily because of the time zone differences. Rarieya used a digital camera to take pictures of anonymous AIDS orphans and families affected by the AIDS epidemic. She uploaded the pictures to RVDS so as to share them with the U.S. team. She also used a combination printer-scanner, which she brought with her to Kenya, to scan the sketches done by participants; again, she uploaded these to RVDS for the RPI team in the United States to access. Finally, Rarieya used a digital camera to take pictures of the final visual campaign graphics. Again, she uploaded them to RVDS for the RPI team to see and critique but not modify.

PARTICIPATORY DESIGN METHODOLOGY

In conducting the collaborative learning workshop, a key issue was how to demystify the design process by developing an adaptable communication design methodology that is collaborative, is user-friendly, and can be made virtually accessible to multicultural and multidisciplinary participants. Our objective, which was to conduct a small-scale workshop with grassroots participants, modeled on the studio design course that Bennett had taught the previous term at RPI, enabled us to adapt the structured methodology used by Bennett's students.

Bennett's methodology for participatory graphic design (Bennett, 2006) guided the Kenyan participants step-by-step through the design process from its conception to the production of 11″ × 17″ posters and 5.5″ × 8.5″ postcards for use in campaigns. Participants were asked to keep a design process journal in order to learn how to systematize, manage, and supervise their creative problem-solving process. At the end of each step in the methodology, each participant was required to solicit and document audience feedback (a participatory and evaluative mechanism) about their work in progress. They documented the feedback and presented it for discussion in individual/group critiques that took place indirectly with the RPI professors within the virtual design studio and face-to-face with Rarieya and the other Kenyan workshop participants at the community center.

The design process was carried out in the following phases:

- *Phase I.* The first phase entailed an all-encompassing effort to define the problem, conduct research, and gather audience input. The participants researched HIV/AIDS awareness and prevention in their local communities and discussed their own experiences with the disease. Some of the participants were HIV positive and shared personal stories of how they were infected.

- *Phase II.* In the second phase, the participants moved on to conception, which includes generating ideas, copywriting, and acquiring audience input. They conceived, as a group, several unrelated concepts and tag lines to present to their Kenyan peers for feedback, which they documented in design process journals via Microsoft Word under Rarieya's supervision. After acquiring feedback from the target audience, they moved forward by eliminating the weakest concept.
- *Phase III.* In this phase, participants moved on to visualizing their ideas with text- and image-based graphics in thumbnail sketches of the final layout. Audience feedback was once again required. After considering the audience's suggestions, the participants moved forward by, once again, eliminating the weakest concept.
- *Phase IV.* In this final phase, the participants digitized their hand-drawn thumbnail sketches for the purpose of documenting their process (in the journals) as well as for the purpose of rendering the posters and postcards using real image-based graphics and typesetting. Rarieya introduced the Kenyan lay participants to the industry-standard design software applications for image editing (i.e., Adobe Photoshop), illustration (i.e., Adobe Illustrator), and page layout (e.g., Adobe InDesign).

ASSESSING THE OUTCOME OF THE VDS

In the GCP project mentioned earlier in this chapter, the Russian students were taking the course to satisfy an English language requirement. Thus, everyone conveniently spoke English. Our project might not have been viable without a Kenyan mediator—an English speaker who had knowledge of the Luo tongue of Kenya and of Kenya's NGOs and sociopolitical infrastructure. Having both an on-site English- and Luo-speaking graduate student—a native of Kenya—and a structured design methodology provided crucial resources for intercultural collaborative design. Rarieya made the project viable by assisting the Kenyan laypeople in tracking and documenting their design processes, keeping journals, and creating the campaign. Her expertise in Kenyan culture and the structured methodology enabled her to easily guide the Kenyan lay participants through the visual design of new HIV/AIDS awareness and prevention messages and, in a way, through the redesign of their existing intervention messages—like those in Figure 1—as more effective cross-cultural graphics—like that in Figure 3.

The effectiveness of the graphics in the poster for a Kenyan audience shown in Figure 3 is measured in part by the participatory process undertaken to design it rather than solely by established heuristics for effective cross-cultural graphics developed by design and technical communication professionals within the past century (including Bringhurst, 1992; Clair & Busic-Snyder, 2005; Coe, 2001; Faigley, George, Palchik, & Selfe, 2004; Frascara, 2004; Kostelnick & Roberts,

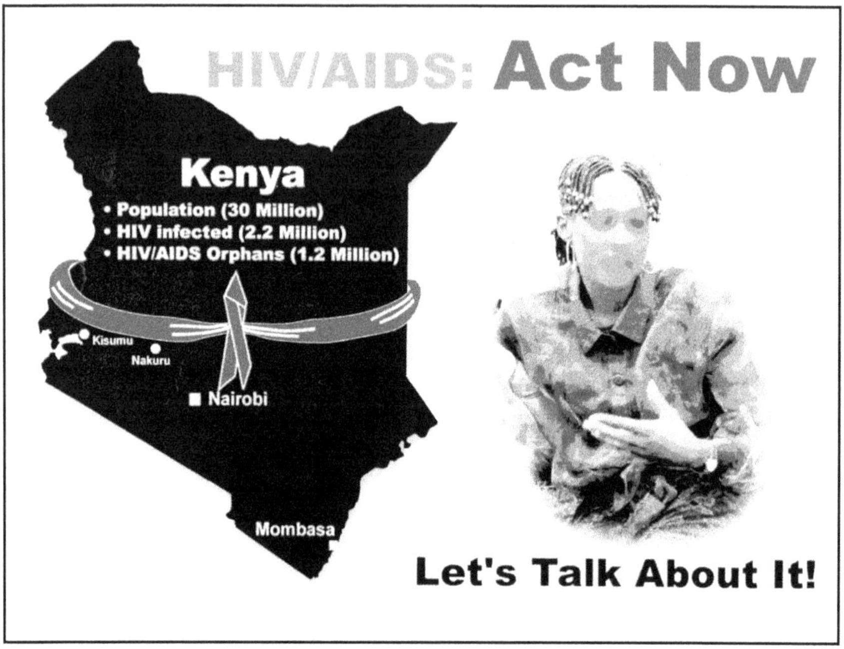

Figure 3. Example of campaign graphics designed by Kenyans. The face in the image on the right has been disguised at the subject's request. It was not censored in the local version of the design.

1998; Lipton, 2002; Lupton, 2005; Meggs, 1992; Schriver, 1997; Tschichold, 1998/1928). In our opinion, the graphic in Figure 3 is more effective communicatively than those in Figure 1 because of its use of more eye-catching and memorable image-based graphics. It is also more effective cross-culturally because it was designed by Kenyan laypeople. Thus, we anticipate it will resonate well with a broader Kenyan audience.

In terms of cross-cultural visual aesthetics, one of the most important outcomes of the workshop was the use of proverbs by Kenyan laypeople. This began with one of the first of the Kenyan sketches, titled "AIDS: Fight the enemy." In e-mail conversations with Rarieya, we pointed out the potential this slogan had to encourage violence against infected individuals. We asked Rarieya to solicit her group (and herself) for some local proverbs that might be relevant to AIDS issues. These included the following:

- *The disobedient fowl obeys in a pot of soup.* (originally from Nigeria)
- *The end of an ox is beef and the end of a lie is grief.*

- *It takes a whole village to raise a child.* (originally from Nigeria; used here to reference African children who are orphans because their parents have both died from AIDS)
- *A snake is in the gourd.* (see Figure 6 for an explanation)
- *A bull dies with grass in its mouth.* (meaning: Live fast. Die young.)
- *All that glitters is not gold.*

The last seemed disappointing to us, as it was a familiar European proverb. The Kenyans, however, replied that it was indigenous, and we began to doubt our initial certitude (as international comparisons of proverbs often show parallel development). Certainly, it was readily put to use by the Kenyans as the graphic in Figure 4, which visualizes their view of AIDS infection as the result of a Western-inspired lifestyle of bars, alcohol, and promiscuous sexuality. An even stronger claim for the value of indigenous life style is oddly visualized in Figure 5. We say oddly because the topless clothing style is nowhere used in the area in which the participants live. The proverb here cleverly answers our critique of the "Fight the enemy" proverb (see Figure 5, caption).

Figure 4. Example of campaign poster/postcard graphics designed by Kenyans. This example uses an African proverb, *Jaber jahula* (English translation: All that glitters is not gold.)

Figure 5. Final example of campaign poster/postcard graphics designed by Kenyans. The proverb cited here implies that AIDS is like a snake in a gourd: You cannot solve the problem by breaking the gourd because it is valuable; likewise, you cannot solve the AIDS crisis by attacking people who are infected. The omniscient mouse relays the message of the poster (this is probably written in English because the message comes from the West)—it is the internationally known "ABC" approach to HIV prevention—abstaining from sexual intercourse, being faithful, and consistently using condoms (hence the acronym "ABC").

Technical Limitations and Possibilities

In our VDS design, we encountered many interesting computer science problems that need further research. Some of the problems we encountered include controlling spam in newsgroups, increasing the bandwidth through compression techniques, and communicating through inexpensive mobile devices. Constructed out of existing technologies, the VDS had some major limitations that affected the quality of the collaborative exchange. It was not the virtual design studio we envisioned. For instance, multiple participants could view a sketch but not modify it simultaneously in real time or archive prototypical stages of the design process. Some of the software applications malfunctioned during critical moments. We were unable to open Adobe Photoshop in Learnlinc during a collaborative session. Limitations in memory prevented the showing of multiple views of images in Yahoo. Also, as evidenced during the preliminary testing of Yahoo within Bennett's studio design course, the chat rooms in Yahoo did not allow for the archiving of group critiques. We had to awkwardly take low-resolution screen shots of the chats and store them on our hard drive. In Bennett's studio design course, it had been determined that the designer's retention of audience feedback was a critical factor in a user-centered design process. Thus, in Kenya, Rarieya helped participants to document—as best they could—audience feedback throughout the design process journals.

A key set of issues that resulted from the VDS workshop involved questions about the possibilities for innovative direct use of instructional and communication technologies. For example, how can we technically enable access by Kenyans to the virtual design studio at multiple field sites, in the current digital divide situation, with only one computer for every 200 (or more) citizens (Mungai, 2002). What technological strategies can be used to compensate for the relatively low density of computers in Kenya? For example, are there new methods of using (liquid crystal display (LCD) projection to expand beyond the single-user assumption? Could the use of WebTV, which received such a lukewarm reception in the United States, play an important role in a society with radically different communication traditions and technological access? Are there ways to recycle TV screens, cell phones, or palm pilots from the United States? The design of rural communication networks, for example, which is facilitated by cell phones, personal digital assistants (PDAs), and kiosks, could also be facilitated through the VDS we envisioned at the start of the project. Communication can be critical in supporting women's rights or collective action, in providing information hotlines, and so forth. Such networks would also be an ideal location for examining new possibilities for hybrids. The enormous number of used cell phones in the United States, for example, is one possible source for the importation of low-cost, high-impact communication technologies. Rarieya observed that one specific cause of the spread of AIDS in Kenya is the requirement that widows marry their husband's brother. If we can find a role that technology might play as an intervention in this process—setting up a women's rights information hotline or facilitating more informed communication networks among women—it would be a powerful contribution to interdisciplinary discourse on global AIDS activism.

Budget Limitations

By replicating Bennett's studio design process in a VDS model linked to a local community computing center in Kenya, our team trained Kenyan participants in the use of Bennett's user-centered methodology as a framework for collaborative learning in which the participants designed their own AIDS prevention communication poster and postcard graphics using a Kenyan visual vernacular. While Bennett's communication design course serves as a practical, directly implemented model of how an AIDS/HIV awareness campaign could be created through collaboration with Kenyan laypeople, we lacked the resources to mass produce the graphic designs or to implement the campaign and measure its effectiveness in decreasing HIV infection in Kenya. Specifically, a lack of resources inhibited us from doing the following:

- *Exploring new technical approaches to participatory and user-centered design for HIV/AIDS awareness and prevention in Africa.* For instance, there was no opportunity for us to create new communication paradigms between folk communication methods and high-tech ones like chat rooms, bulletin boards, and E-mail.
- *Developing a full-scale VDS.* Our preliminary findings reveal that the workshop could profitably be repeated on a larger scale with more Kenyan grassroots participants synchronously interacting with American students and faculty. The full-scale VDS would enable multiple participants to view and modify a sketch synchronously, to simultaneously post comments for the rest of the participants to see and critique, and to archive the prototypical stages of a design project and the collaborative critiques of the creative development of the project. The VDS we envision would accommodate no more than 25 (the same limit for student enrollment in Bennett's visual communication course) local and remote participants, would be accessible and fully functional via a computing center in Kenya with low-bandwidth Internet access, and would offer the following interactive features within its interface:
  - Combined image-editing, drawing, and layout design tools
  - An enhanced whiteboard that is more user-friendly for grassroots people and more economical than current industry-standard graphics software and that enables multiple participants to view, critique (verbally and visually), and modify a multimedia composition synchronously and asynchronously

- An interface that is more easily adopted by Kenyan participants (after basic computer training has been conducted) through its culturally appropriate visual coding
- A storage mechanism that enables multiple sessions in a design activity
- A history feature that logs daily activity within the VDS

Another important VDS component that could at least have been explored with adequate budgetary resources is the use of the existing East African Bao game, also known as Ajua (in Kenya) and Mancala (in the United States), for communicating HIV/AIDS awareness and prevention information. In Kenya (and elsewhere in Africa), this traditional pebble or seed board game is widely played as a recreational activity. The game design and rules vary from one cultural group to another. Traditionally, a typical Bao game involves a wooden board, carved out of hardwood, with a total of 16 hollows, carved in two rows, and 48 small round seeds (or other small objects) for pieces, as depicted in Figure 6. Normally, there are two extra hollows, which are centrally placed at the ends of the rows.

Figure 6. Ajua board game.

English	Luo
foot	*tielo*
hip area[a] (can be associated with mates and infections)	*je, sundi,* or *yunga*
thigh	*bam*
eye[a]	*wang*
chest	*kor*
shoulder	*nyiwi, gok,* or *agoge*
neck	*ngudi*
head	*wich*

[a]These are important game areas that could relate to communicating HIV/AIDS awareness and prevention.

Figure 7. Names of Bao pits and their English translations.

The game is played with 48 seeds on a rectangular board containing 16 receptacles, arranged in pairs, as noted above. The players, usually two, sit with the board crossways between them. At the start of the game, 3 seeds are placed in each receptacle, or house. At the beginning of the game each of the playing pits should contain 3 seeds. Placing 3 seeds in the pit initially just confirms that each player has 24 seeds at the beginning of the game. Each seed in the pit is then placed, one seed at a time, in the successive pits, moving clockwise around the board. A player starts each of his moves from a pit on his side and moves clockwise to the opponent's pits, dropping one seed during each move. From the interviews that Rarieya conducted near Kusa, we discovered that the current convention for naming the pits could be utilized for our HIV/AIDS project, in that it represents parts of human body (i.e., the female body) as noted in Figure 7.

Today, the game is played by young men who modify the game rules. In the past, it was typically used only for entertainment purposes. Now, the Bao game is used both for entertainment and for economic gain as players gamble for amounts of money ranging from hundreds to thousands of Kenya shillings. We propose that it could also be appropriated for health communications.

CONCLUSION

Each discipline made its own contribution to the project, but it was the synthesis of the three that offered a solid foundation for understanding the cultural and technical issues that we faced and that future technical communicators will face in online learning environments intended for first- and third-world access. This type of contribution to interdisciplinary discourse is valuable even in a first-world context, when the communication has to take place across different cultural groups. In our future work in this area, we envision the participation of other experts as well—in fields such as public health or medical anthropology, for example. The progress we've made so far with the VDS in a third-world context moves us further along in deriving participatory design paradigms for international online learning and health communication between first- and third-world participants.

REFERENCES

Andrews, D. C. (2000). A participatory approach to communication for developing countries . In P. H. Hager & H. J. Scheiber (Eds.), *Managing global communication in science and technology* (pp. 67–83). New York: John Wiley & Sons.

Barabasi, A. (2002). *Linked: The new science of networks.* Cambridge: Perseus.

Bennett, A. (Ed.). (2006). *Design studies: Theory and research in graphic design.* New York: Princeton Architectural Press.

Beyer, H., & Holtzblatt, K. (1998). *Contextualdesign: Defining customer-centered systems.* San Francisco: Morgan Kaufmann.

Bringhurst, R. (1992). *The elements of typographic style.* Vancouver, BC, Canada: Hartley & Marks.

Cherkasky, T., Greenbaum, Jr., Mambrey, P., & Pors, J. K. (Eds.). (2000). *Designing digital environments: Bringing in more voices: Proceedings of the Participatory Design Conference, November 2000, CUNY.* New York: CPSR.

Choi, K. H., & Coates, T. J. (1994). Prevention of HIV infection. *AIDS, 8*(10), 1371–1389.

Clair, K., & Busic-Snyder, C. (2005). *A typographic workbook: A primer to history, techniques, and artistry.* New York: John Wiley & Sons.

Coe, M. (2001). *Human factors for technical communicators.* New York: John Wiley & Sons.

Dezso, A., & Barabasi, A. (2002). Halting viruses in scale-free networks. *Physical Review, 65,* 055103-1, 055103-2, 055103-3, 055103-4.

Downey, G. L., & Lucena, J. (1997). Engineering selves: Hiring into a contested field of education. In G. L. Downey & J. Dumit (Eds.), *Cyborgs and citadels: Anthropological interventions in emerging sciences and technologies* (pp. 117-141). Santa Fe, NM: School of American Research Press.

Eglash, R. (1999). *African fractals: Modern computing and indigenous design.* New Brunswick, NJ: Rutgers University Press.

Eglash, R., Bennett, A., O'Donnell, C., Jennings, S., & Cintorino, M. (2006). Culturally situated design tools: Ethnocomputing from field site to classroom. *American Anthropologist, 108*(2), 347–362.

Faigley, L., George, D., Palchik, A., & Selfe, C. (2004). *Picturing texts.* New York: W. W. Norton.

Forslund, C. J. (1996). Analyzing pictorial messages across cultures. In D. C. Andrews (Ed.), *International dimensions of technical communication* (pp. 45–58). Arlington, VA: Society for Technical Communication.

Frascara, J. (1997). *User-centered graphic design: Mass communications and social change.* Bristol, England: Taylor & Francis.

Frascara, J. (2004). *Communication design: Principles, methods, and practice.* New York: Allworth Press.

Geary, D. C. (1994). *Children's mathematical development: Research and practical applications.* Washington, DC: American Psychological Association.

Giammona, B. (2004). The future of technical communication: How innovation, technology, information management, and other forces are shaping the future of the profession. *Technical Communication, 51*(3), 349–366.

Hager, P. J. (2000). Global graphics: Effectively managing visual rhetoric for international audiences. In P. J. Hager & H. J. Scheiber (Eds.), *Managing global communication in science and technology* (pp. 21–43). New York: John Wiley & Sons.

Herrington, T., & Tretyakov, Y. (2005). The Global Classroom Project: Troublemaking and troubleshooting. In K. C. Cook & K. Grant-Davie (Eds.), *Online education: Global questions, local answers* (pp. 267–283). Amityville, NY: Baywood.

Horton, W. (2005). Graphics: The not quite universal language. In N. Aykin (Ed.), *Usability and internationalization of information technology* (pp. 157–187). Mahwah, NJ: Erlbaum.

Kiwanuka-Tondo, J., & Synder, L. (2002). The influence of organizational characteristics and campaign design elements on communication campaign quality: Evidence from 91 Ugandan AIDS campaigns. *Journal of Health Communication, 7*(1), 59–77.

Kostelnick, C., & Roberts, D. D. (1998). *Designing visual language: Strategies for professional communicators* Boston: Allyn and Bacon.

Lipton, R. (2002). *Designing across cultures: How to create effective graphics for diverse ethnic groups*. Cincinnati, OH: How Design Books.

Lupton, E. (2005). *Thinking with type*. New York: Princeton Architectural Press.

Meggs, P. (1992). *Type and image: The language of graphic design.* New York: John Wiley & Sons.

Mungai, W. (2002). The African Internet: Impact, winners and losers (a paper on the impact of the internet on human development in Africa). Retrieved September 25, 2002, from http://www.geocities.com/wainainam

Ogbu, J. (1998). Voluntary and involuntary minorities: A cultural-ecological theory of school performance. *Anthropology and Education Quarterly, 29*(2), 155–188.

Schriver, K. A. (1997). *Dynamics in document design.* New York: John Wiley & Sons.

Schuler, D., & Namioka, A. (1993). *Participatory design: Principles and practices.* Hillsdale, NJ: Erlbaum.

Scrivener, S. A. R., Ball, L. J., & Woodcock, A. (Eds.). (2000). *Collaborative design: Proceedings of the Co-Designing Conference, Coventry School of Art and Design.* Coventry, England: Springer.

Tschichold, J. (1998). *The new typography.* ®. McLean, Trans.). Berkeley: University of California Press. (Original work published 1928.)

Wojtowicz, J. (Ed.). (1995). *Virtual design studio.* Hong Kong: University of Hong Kong Press.

CHAPTER 9

Cultural Adaptation of Cybereducation

Judith B. Strother

CHAPTER OVERVIEW

When a Web-based training program is being conceptualized, cultural implications must be considered during all phrases of development—planning the course curriculum, designing the user interface, and delivering both the online and on-site portions of the program when a blended learning model is used. This chapter uses a case study of a Web-based "aviation English" course to illustrate the complex and multifaceted aspects of culture—national, professional, and pedagogical—that can influence the way the training is designed and delivered and the way the trainees interact with all facets of the course.

Cultural issues pose an interesting and important set of considerations when one is attempting to design online training materials for an international audience. At a fundamental level, the question is, "How do cultural differences affect online learning?" At an applied pedagogical level, we must ask, "Can virtual learning environments be designed to accommodate the needs and facilitate the learning of a culturally diverse audience, and if so, what is the best methodology to accomplish this?"

Any kind of cultural adaptation must go beyond the usual national cultural descriptors and deal with the often complex subcultures that can have a significant impact on trainees' interactions with the Web-based training (WBT) program—both technologically and pedagogically. For an international WBT program, there are three main areas in which culture must be considered. First, when program designers are developing all aspects of course content, culture must be a significant aspect of audience analysis. Second, when they are designing the user interface, the need for both cultural and linguistic adaptation must be analyzed. Third, when Web-based training is delivered in a blended learning model, consideration of cultural pedagogical preferences plays an essential role with regard to the on-site portion of the curriculum.

This chapter examines the multiple layers of cultures and relevant subcultures—national, professional, and pedagogical—that can affect a group of trainees. The challenge of being sensitive to cultural issues while developing Web-based training programs for a global audience becomes greater when there are compelling reasons for standardizing the course content for all trainees. This chapter presents a case study of the development and delivery of online training in aviation English for an international audience of commercial pilots, where there are regulatory constraints on adapting the content to local cultures. This raises a very real and practical distinction between globalization (where commercial or regulatory requirements may predominate) and localization (where additional pedagogical issues should be considered). Under these more restrictive conditions, the question now becomes, "How does an online training program cope with the need for international standardization and the reality that localization may be, from a pedagogical standpoint, a more effective way to deliver significant parts of the product?" In this case, cultural adaptation is mainly accomplished in the on-site portion of the blended learning curriculum, not in the online portion of the product.

Some educators distinguish between teaching and training, with the term *teaching* being used more for higher-level cognitive or knowledge-based activities and *training* being used more frequently for skill-based tasks. In fact, for both teaching and training, the instructor and/or course delivery system have similar goals or learning objectives—to do whatever it takes to ensure that the student/trainee is able to learn and apply the material delivered. Within the corporate world, the term "training" is usually used for any kind of educational activity. Therefore, in the case study used in this chapter, while teaching would be a more common term for a language course, the term training is used because the course is usually delivered within a corporate framework.

CULTURAL FRAMEWORK

While culture is a complex abstraction, there are many practical cultural issues that must be identified and understood in order to succeed in developing and delivering e-learning or blended learning courses to a global audience. To use Hofstede's (1991, p. 5) classic definition, culture is the "collective programming of the mind which distinguishes the members of one group or category of people from another." An important construct for this chapter is the fact that culture can act as a mediator between the universal characteristics of humanity and the personality characteristics of an individual. Within a specific culture, there tend to be clusters of personality types (Rosman & Rubel, 1995), which help us make generalizations about the values and behaviors of a specific cultural group.

National Culture

A number of researchers have produced ways of categorizing cultural dimensions. Some have a certain focus or use a specific professional framework (e.g., House, Hanges, Javidan, Dorfman, & Gupta, 2004, with the Global Leadership and Organizational Behavior Effectiveness—GLOBE—study; and Schwartz, 2004). Most have been based on or correlated with the classic model presented by Hofstede (1991, 2001; Hofstede & Hofstede, 2005) which can be summarized as follows:

- *Power distance*—The level or degree of inequality that individuals within a society accept or allow to exist between themselves and those in authority.
- *Uncertainty avoidance*—The extent to which individuals will go to avoid uncertainty in their lives, which is related to the level of stress in a society in the face of an unknown future.
- *Individualism versus collectivism*—The degree to which individuals are integrated into groups. In individualistic societies, individuals look after themselves and their families, whereas in collectivist societies, people are integrated into strong, cohesive in-groups, which protect them in exchange for unquestioning loyalty.
- *Masculinity versus femininity*—The division of emotional roles between men and women, which is related to the extent to which people are assertive and competitive.
- *Long-term versus short-term orientation*—The choice of focus for people's efforts: the future or the present and past.

However, Hofstede provides an important caveat, or what he calls a "pitfall," about confusing national culture with other levels of culture. He points out that his "dimensions were chosen so as to discriminate among national and maybe regional and ethnic cultures, but not for other distinctions like gender, generation, social class or organization (Power distance, exceptionally, is also relevant for social classes, occupation, and education levels; masculinity is relevant for distinguishing occupations)" (Hofstede, 2002).

Triandis (2004) and Triandis and Tratimow (2001) have dealt with seemingly contradictory characteristics within the same individual. For example, a person could rate high on both the collectivism and the individualism scales—a surprising, if not contradictory finding. A study that is important for our case population of commercial pilots who fly international routes is that of Yamada and Singelis (1999), which has shown that a person who was raised in a collectivist culture and then lived in an individualist culture for several years might rank high on both individualism and collectivism.

Hall (1976) originally proposed the concepts of *high context* and *low context* related to cultures (see also Hall & Hall, 1990). In a high-context culture, more information is embedded in the communication context, which is shared by everyone in the group. Therefore, people gather information from their environment rather than from explicit information given in the communication itself. In a low-context culture, where little common background information is shared by the collective group, communication tends to be longer, wordier, and more explicit. Typically, cultures that are collectivist tend also to be high-context (for example, most Asian countries) and those that are individualistic tend to be low-context (for example, the United States). Individuals from high-context cultures and low-context cultures have different perspectives on and approaches to education, so this is one additional aspect that must be considered when designing any kind of cybereducational program (see Morse, 2003, for an overview).

Schwartz and Bardi (2001, p. 287) suggest taking an interesting counter-perspective to achieve new insights—a similarities perspective, focusing on "the largely shared, pan-cultural value hierarchy [that] lies hidden behind the striking value differences." Theirs is a valid perspective because, to fully understand the priorities that drive human behavior, we need to understand both the similarities and the differences and the interactions between them.

Professional Culture

Within any national culture, a number of subcultures exist. For this case study, the relevant professional subculture is that of commercial pilots who fly international routes.

A group of researchers at the University of Texas (Helmreich, 2000a, 2000b; Helmreich & Merritt, 1998; Merritt, 2000) conducted an extensive aviation-specific study based on new measures of culture that incorporate Hofstede's original dimensions from his survey of IBM personnel (Hofstede, 1980). Helmreich and Merritt surveyed more than 8,000 pilots in 26 countries.

Helmreich and Merritt described three intersecting cultures that surround every flight crew—national, organizational, and professional. Two of Hofstede's dimensions of national culture, discussed earlier, are particularly important in the cockpit—Individualism/Collectivism and Power Distance. When translated into individual characteristics, they especially influence the ways in which juniors relate to seniors. In societies having a low power distance ranking, subordinates feel closer to those in charge and therefore are more likely to question an instruction or order they believe is wrong or inappropriate. On the other hand, when power distance is high, subordinates are more likely to obey orders without questioning the validity or correctness of the order. For example, a junior member of the cockpit crew may be hesitant to question or challenge a senior officer's order or decision even when he or she knows it to be in error, or he may be hesitant to speak up and provide critical information if it appears to contradict a senior officer's order. The great reluctance of subordinates in the military, aviation, and medical professions to challenge authority figures is deeply ingrained in and reinforced by their professional and organizational cultures as well as their national cultures.

Organizational culture (often called corporate culture) may operate within a national border or transnationally. The corporate culture of the airline influences openness of communication, attitudes toward human error, and level of trust between management and flight crew. It will also have a strong influence on attitudes toward training, for example, how training is prioritized and validated within the organization.

Professional culture has a strong presence in the aviation industry. Historically, the captain of a ship or an aircraft has complete authority and total responsibility for the operation and safety of his vessel. This principle has traditionally formed the backbone of a strong professional culture among both maritime and aviation crews.

There is a high level of professional pride in being a pilot, a healthy respect for others in this elite career group, and a strong motivation to perform well. The validity of this culture of pilots is nowhere better illustrated than during World War I, when pilots on both sides of the conflict often viewed their adversaries with a strange sense of kinship and respect. While this fraternity of pilots is intangible, it nevertheless exists in a very real sense and is based on the shared, common experiences of the community of pilots. Along with that goes a macho sense of invulnerability and a failure to rely on fellow crew members. When this overlaps with the power distance characteristic, it can create a dangerous situation.

In aviation particularly, bad decisions or commands based on flawed information can quickly lead to disaster. Therefore, it was with safety in mind that the CRM (cockpit resource management) concept was introduced. When CRM operates as it should, any member of the cockpit flight crew, regardless of rank or status, is encouraged and empowered to speak out if he or she feels that the safety of the aircraft may be compromised by a given command or instruction. This "safety culture" becomes part of aviation's professional culture, but its implementation is still influenced by national culture.

Undoubtedly, the national, professional, and organizational cultural values are interconnected in complex ways. In any event, when they all support the same point of view or philosophy, then the belief structure will be reinforced. This is why many international airlines have had difficulty implementing a truly effective CRM program. The challenge of implementing an effective CRM program across national boundaries shows how even beneficial programs can falter due to cultural biases. It also shows how, with little extrapolation, the effectiveness of training, whether delivered strictly online or through some other model, can be compromised or diminished, especially if the program developers and/or instructors do not understand these types of cultural differences and intercultural conflicts.

While studies such as that of Yong (2003) have found that, in addition to national culture, organizational culture has a strong influence on training practices, Helmreich and Merritt found that, while professional and organizational cultures do have an impact on flight operations, national culture has a much stronger influence. It has also been shown that culture is strongly associated with a preference for automation and with regard to the way it is used (Sherman, Helmreich, & Merritt, 1997). This finding holds particular significance for online training program development and delivery.

Within the professional pilot culture, other cultural influences include military versus commercial organizational culture and the influences of the culture in which the ab initio training was carried out versus the home culture. As mentioned earlier, when people spend time away from their home cultures, they may take on characteristics of the host culture. However, the national culture tends to exert the strongest influence (Yamada & Singelis, 1999). For example, an Asian pilot may have been trained in a Western country such as the United States. However, when that pilot returns to Asia to fly for a commercial airline, the home country's culture of practices may conflict with the Western teachings and thereby affect the cockpit culture of that pilot and his or her crew. When designing online training materials for this specialized audience, one needs to examine the cultural elements that might be expected to exert the strongest influence on the effectiveness of the training.

Pedagogical Culture

Another aspect of culture that is relevant to this case study is the aspect that frames the educational system of a group of people—namely, the pedagogical culture, which incorporates the organizational structure of education within a particular society. While we must recognize that there are wide variations in the individual characteristics and preferences of both learners and teachers within the same culture, we can generalize about certain characteristics and behaviors that are typical of national cultures or various subcultures.

Cultural biases about appropriate educational models run very deep and influence the way members of a particular culture act and feel. Therefore, it is important to understand these cultural biases before developing training materials and programs to be used internationally, whether the materials are to be used strictly online or in some combination of online and site-based delivery, that is, in a blended learning model.

With these views in mind, we need to examine how national cultural dimensions affect learner attitudes, since culture implicitly affects the way in which training is presented and learning is accomplished. According to Planel (1997), students' attitudes to education are derived from their sociocultural backgrounds, and students interpret both what and how they learn through the medium of the culture to which they belong. A student's understanding of such educational values as authority, thought, control over learning, and educational goals and how to achieve them is related to national culture and thus has an effect on student motivation. Planel further suggests that cultural values are more significant for learning than pedagogical styles and argues that underlying educational values give meaning to styles of pedagogy. With that said, it is essential to analyze pedagogical models to get a handle on differences in instructional expectations such as instructor and student roles, academic norms, and acceptable classroom behavior.

To take this a step further, in the design of any form of cybereducation, we have to realize that the effects of information and communication technologies are often mediated by the learner, the content, the instructor, and the learning environment (see, e.g., Zhao & Seppo, 2002).

A variety of pedagogical models exist and there are numerous ways to label them. For example, Borytko (1991) proposed three paradigms of pedagogical philosophy or thought. The first is the sciento-technocratic paradigm in which knowledge and information—not the student—are at the center. Therefore, lack of knowledge is equated with inadequacy or lack of power. The teacher is in control. The second paradigm is the humanitarian paradigm. In this model, which emphasizes experiential or activity-based learning, the mode of knowledge acquisition is considered to be more important than the knowledge itself. The esoteric paradigm is the third model of pedagogical philosophy. It focuses on participation in relationships, that is, emotion and value-based experience in which the student is to develop an attitude and a set of values.

Using this particular framework to examine national pedagogical cultures, we find that a number of cultures have a preference for the sciento-technocratic paradigm, or what we in the Western world would call a teacher-centered model. In contrast, many U.S.-based educational institutions use the learner-centered model, which correlates more closely with the esoteric paradigm.

Although considering national and pedagogical cultures is essential when designing any kind of cybereducation, course designers need to be cognizant of the fact that there are a myriad of individual characteristics that must be taken into account. Individual students tend to prefer one learning style to another and incorporate multiple intelligences in their learning process (Strother & Alford, 2003). These learner variables are set within national preferences for training/learning styles. There is a growing body of research that illuminates the interaction of national culture, learning styles, and e-learning experiences (e.g., Banham & Wong, 2001; Cossel-Rice et al., 2001; Niles, 2005; Payne, 2002; Ramburuth, 2000; Strother, 2003a, 2003b; Strother, Chiu, Serrano, & Alford, 2005; Volet & Renshaw, 1996).

The primary question is as follows: What part does each facet of culture—national, professional, and pedagogical—play in the development and implementation of a Web-based curriculum designed for a multicultural audience? While culture must be considered throughout the process, usually it does not play an equal role in each phase of the program's creation: development of course content, design of user interface, and implementation of the on-site part of a blended learning model. These factors will be addressed through a case study of a Web-based language training program that was developed for a specialized group of international trainees.

CASE STUDY: WEB-BASED AVIATION ENGLISH

An English for specific purposes (ESP) course—a dual-purpose course involving specialized content along with foreign or second language instruction—is a challenging course to develop under any conditions. When the course is designed within a blended learning model—Web-based and site-based instruction in this case—for international distribution to a very specialized group of trainees, additional complexities emerge. In this case study, multiple cultural and subcultural factors overlap—national, professional, organizational, and pedagogical. Added to these are individual learner variables, which are also set within national preferences for training/learning styles.

In this particular case study, the course content is designed to teach English as a foreign language, more specifically, English for specific purposes—here, aviation English. Within this highly specialized field, there are multiple constraints on the course content. The first is the topic-restricted aviation content itself. The next is the professional skills of the trainees, who are commercial pilots who fly international routes and who are based in member states of the United Nations. As already discussed, they bring with them multiple layers of professional and organizational culture in addition to the culture of their national origin. In addition to demonstrating fundamental knowledge within the field of aeronautics and the specific knowledge required for a commercial pilot's license, this group of trainees must prove they have a high enough level of English-enough fluency, both in general English and in the specific skill area of radiotelephony, at the end of the course. The proof will be an internationally standardized aviation English proficiency examination. So this cybereducation course must prepare them for a high-stakes exam within a very specific skill area while providing a solid foundation of English language proficiency.

Florida Institute of Technology (Florida Tech) uses the aviation English product of Aviation English Training, Inc. (AET) to deliver Web-based language training at four proficiency levels. The course design incorporates a patented learning retention process from Profound Learning, Inc., which continuously adapts to each trainee's learning needs. This adaptive system remembers what each trainee learns and does not learn and then customizes a study plan that is unique to each trainee.

Strictly Online versus Blended Learning Delivery

Although the program can be completed totally online, Florida Tech highly recommends a pedagogical model that takes a blended learning approach because of its strong advantages. Studies now empirically demonstrate the benefits of a blended learning model. Rovani and Jordan (2004) and Heinze and Procter (2004) have demonstrated that blended learning produces a stronger sense of community, which facilitates better learning processes. In addition, research shows that blended learning courses yield success rates equal to, and in many cases greater than, their fully online or classroom-based counterparts; they are usually preferred by students; and they enjoy a higher retention rate (Buzzetto-More & Sweat-Guy, 2006; Dziuban, Hartman, & Moskal, 2004; Strother, Chiu, Serrano, & Alford, 2005). Blended learning courses can also provide a combination of student-centered and traditional teacher-centered learning (e.g., Lanham & Zhou, 2003), which is especially important in transnational cyber-education programs. As discussed later in the chapter, acculturated blended learning can help bridge the mismatch between the different pedagogical paradigms of trainers and trainees.

> The power of blended learning is that it integrates seemingly opposite approaches, such as formal **and** informal learning, face-to-face **and** online experiences, directed paths **and** reliance on self-direction, and digital references **and** collegial connections, in order to achieve individual *and* organizational goals. (Rossett & Franzee, 2006)

This definition captures all the key elements, clearly demonstrating exactly why it is so challenging to consider the culture—whether national or professional—of a blended learning course when designing it.

One of the best practices in blended learning is the identification of components that truly require face-to-face instruction (Shaw & Igneri, 2005). Within an e-learning context, most of the discussions of cyberculture or cultural factors deal with interactions among the course participants—trainer to trainee or trainee to trainee in a variety of media such as chat rooms and instant messaging. In this case study, when clients choose the blended learning approach, Florida Tech brings trainees together for a one- or two-week classroom session at the end of the training at each language proficiency level. Here, interaction among trainees and between trainer and trainees occurs, and trainees receive intensive face-to-face training in the most problematic area for many nonnative English speakers—high-quality oral communication skills. These commercial pilots must be able to communicate clearly using precise radiotelephony procedures, which involve very terse but critical pilot-to-controller communication. Florida Tech provides intensive practice in listening and pronunciation activities that are based on actual tapes of pilot-to-tower communication. These are crucial skills for pilots and have high priority in any aviation English training program. It was deemed to be a better use of time to focus on skills such as these instead of activities such as fostering peer-to-peer conversations online. In addition, Florida Tech chose to integrate e-learning with some traditional classroom sessions so that, in addition to the intensive aural/oral practice, cultural preferences for training styles could be more readily taken into account.

Globalization versus Localization of Program

When this particular Web-based program was conceptualized, decisions had to be made about whether all or parts of the educational product—including both the course content and the user interface—could be globalized or localized, that is, to what extent each portion of the product could be culturally adapted.

Localization involves taking a product and making it linguistically and culturally appropriate to the target locale (country/region and language) where it will be used and sold (LISA, 2002). Wherever possible, a Web site should be localized for the area in which it will be used, allowing for adaptation to national and cultural preferences (see, e.g., St.Amant, 2005, for an overview). However, as some studies have shown, it is challenging even to get agreement on the best way to localize the text-based interface (see, e.g., Onibere, Morgan, Busang, & Mpoeleng, 2001). In other cases, such as the case study reported here, localizing the Web site is often not practical, especially for Web-based training that is designed to be accessed around the globe.

Globalization (sometimes called internationalization) is the process of generalizing a product so that it can handle multiple languages and cultural conventions without the need to redesign. Bourges-Waldegg and Scrivener (1998) combine the concepts of localization and globalization in the following way: "Generally speaking, internationalization concentrates on separating the 'cultural elements'—e.g., character sets—of a product from the rest of it; localization is about adapting those 'cultural elements' for a specific target culture. For simplicity, we shall call this internalize-localise process Culturalisation" (p. 287). These researchers go on to propose a design approach for culturally diverse user groups called Meaning in Mediated Action (MIMA) (Bourges-Waldegg & Scrivener, 2000).

In the aviation English case study reported in this chapter, there is a need to deliver content that is internationally regulated and standardized by the International Civil Aviation Organization (ICAO) as well as by other national aviation regulatory bodies, such as the Federal Aviation Authority (FAA) in the United States. Since this content forms the basis for the online part of the course, we can say that the online product has been globalized—the *content* itself is considered "culturally neutral" and can be delivered to all clients around the world without regional modifications. Thus, there was little cultural adaptation of the user interface in this program. The main adaptation of the program Web site itself was linguistic, in recognition of the fact that almost all users of the Web site would be from a wide variety of linguistic backgrounds and would be reading the Web content in English as a foreign or second language. The Web site was fully beta tested with international trainees from a variety of countries, to ensure that the Web site navigation and all representations and program instructions were clear.

In some situations, once an educational institution develops globalized course content deemed suitable for all international learners, translation of that content for specific language groups may be a good choice to make the content more accessible. However, in many cases, translating the interface into multiple languages is not possible. Kukulska-Hulme (2000) presents guidelines for categorizing and dealing with language-related problems, which help both native and nonnative speakers.

In the case study presented here, translation is not an option. The course is an English language training program, so all of the communication and course content must be in English. The Web site has been carefully designed with international markets in mind; however, it has not been localized for multiple specific markets. On the other hand, the site-based classroom session part can be localized for each major region, although it is a complex process. In addition, other kinds of localization can be implemented. For example, marketing efforts must be localized when selling the training program in various countries.

If an educational institution such as Florida Tech decides to localize the site-based part of its aviation English program, what is the best way to accomplish this? As discussed earlier, one of the most challenging parts of designing any cybereducation program to be delivered internationally is the incorporation of the cultural element. The cultural issues are enormous when an educational institution or training company, which is based in one country, needs to deliver classroom sessions in a number of other countries. The educational institution has to be prepared to deal with the mismatch between trainers who probably come from one cultural pedagogical framework (e.g., a U.S. trainer from a learner-centered, individualistic, interactive, communicative pedagogical culture) and a roomful of trainees who come from a different culture (e.g., Chinese, teacher-centered, collectivistic, following a pattern of sitting and listening and memorizing what the teacher says). Consider trainees whose culture teaches them never to disagree with the trainer being trained in an interactive classroom where there are active discussions, even lively disagreements. In some cases, these trainees may even disagree with the trainer. This can create a very uncomfortable and intimidating situation for them.

In addition, think about the individual characteristics of the extrovert versus those of the introvert. "Extraversion scores were higher in individualist cultures" (Hofstede & McCrae, 2004, p. 73). Gudykunst, Yang, and Nishida (1987) earlier found that students from the United States, an individualist culture, score significantly higher on an extraversion scale than students from Japan and South Korea (more collectivist cultures). Extraversion is a trait that comes into play in a communicative classroom, since it is suggested that it helps people initiate contact with strangers and create new social relationships. The trainer should also consider that highly participatory activities are typically more difficult for introverted learners, although such activities are an ideal language learning tool in a learner-centered classroom.

When a trainer is beginning the process of adapting to a particular culture for the classroom sessions of the course, one challenge is to get accurate information about the local culture of the area. It is essential not to overgeneralize but rather to explore all avenues to discover the pedagogical style or styles preferred by a given learner group. Even so, any individual instructor will want to get to know each individual student and let that knowledge guide any decisions that are made about appropriate pedagogical decisions.

In many cases, of course, there is no single pedagogical style for the entire culture. Although China is broadly labeled as being collectivist, teacher-centered, and so forth, it seems to have a number of cultural pedagogical styles. Huang and Leung (2005) have analyzed the apparent paradox of teacher-dominated classrooms: these are assumed not to be conducive to Western-style learning, but Huang and Leung found that Chinese students registered positive achievement levels in these classrooms. They discovered that even in teacher-centered classrooms, there were some activities that could be labeled learner centered according to Western models. While their study focused on mathematics education in Shanghai, their findings may well transfer to classrooms covering other topics and to educational systems in different cultural environments.

Several studies have reported on attempts in international education and training to transition to a learner-centered pedagogy (see, e.g., Tabulawa's work in Botswana, 1998). Tabulawa assumed that the primary reasons for lack of success in introducing a learner-centered pedagogy involved technical and budgetary issues as well as poorly trained teachers. In fact, the teachers tended to make assumptions about the nature of knowledge and how it should be delivered as well as the overall goal of educating their students. Along with possibly faulty perceptions of the students themselves, these assumptions influenced their classroom practices. Tabulawa (2003) later argued that learner-centered models are being forced on schools that receive aid as part of democratization, so the reason for the choice of these models was ideological, rather than academic. However, in our case study, as in many cases, the learner-centered model is not a culturally democratizing choice—it has been determined to be the most successful paradigm for language learning.

Chowdhury (2003) also encountered a mismatch between local and international expectations when teaching English as a foreign language (EFL) in Bangladesh. While this case involved the classroom application of communicative language teaching in a university setting (without an online component), there was a cultural contradiction between the internationally trained teacher and the students, whose cultural expectations were that the teacher should be at the center of their pedagogical paradigm. McCarty (2005), in his work in Japan, discovered the limitations of transferring the western e-learning paradigm into a non-Western culture. Duffy and Osman (2005) found that their students in Azerbaijan were frustrated by their trainers' use of a problem-based inquiry technique since they too were expecting teacher-centered didactic instruction.

These are just a few of the documented experiences that demonstrate the mismatch between the culture of the educators and that of the learners. The overlapping national and pedagogical cultures of the learners framed a set of learner expectations that interfered with learning until adaptations were made.

Not everyone agrees that cultural adaptation is appropriate. Goodfellow and Hewling (2005, p. 356) seem to deny the need to "remedy inequities brought about by the application of pedagogical approaches arising from one cultural context (i.e., social-constructivism in European and North American educational thinking) to groups or individuals whose thinking and behaviour is shaped by wholly different philosophical traditions." However, in their own analysis of two of their graduate education courses, they conclude: "we see the engagement of students from various cultural (in the 'essential' but not exclusively national/ ethnic sense) backgrounds with each of these two overarching pedagogical cultures [teacher centered versus learner centered] as presenting a further challenge to our ability to construct courses which can contribute to the emergence of new 'multi-cultural' narratives for the world's virtual learning environments" (p. 364). This just underscores the complexities of the multiple layers of culture in our learners and trainers and the challenge of how to best deal with them in both the online and the site-based parts of a blended learning program.

No doubt, the desired end result is the same in all these classrooms. Trainers want their trainees to have the best possible learning experience from both the online and the on-site portions of the curriculum. However, trainers and trainees may be defining this end result differently and planning to get to the end result in very different ways.

Choice of Acculturated Blended Learning

In an attempt to formalize the process of dealing with cultural issues for a Web-based course with a multinational on-site component, Strother, Fazal, and Gurevich (2007) have proposed an acculturated blended learning model to handle the need to transition from the nationally preferred pedagogical style to a collaborative learning style for language training within the site-based portion of the course. Their study extends the aviation English case study presented in this chapter, using the example of an on-site session in Russia for the application of the model.

This model makes the broad assumption that the members of a group of trainees all come from the same cultural background and have thus been exposed to the same pedagogical framework. In our case study, we are making the same assumption, although there is the possibility that some of the trainees will have been exposed to Westernized training styles during their flight training. However, they may or may not have been exposed to the learner-centered cooperative learning model that is ideal for language training.

The educational institution employing this model must first learn as much as possible about the general cultural framework and specifically the preferred pedagogical style of delivering training within the trainees' country. Then, a specific strategy has to be adopted to bridge the two pedagogical models/styles and enable the transition to the one that works best for language education.

Another important part of the process is the transition from fully online participation—in which the trainee is in complete control—to a classroom environment in which a shift takes place in almost all aspects of handling of the learning. For the trainees, who are professionals, this is an important consideration. since they have to move from a professional culture of high respect and control into a seemingly subordinated position in a classroom. Lack of language expertise reinforces trainees' perception of this "inferior" position with respect to both the instructor and for many, with respect to their peers.

The acculturated blended learning model includes the following guidelines:

1. Train the trainers: Trainers going into a country to deliver the site-based portion of a blended learning program must be trained in the technical content to be delivered. Then, it is essential that the trainers learn about the national culture, the prevalent characteristics of people, and the preferred pedagogical styles of the culture, in order to operate successfully, arrive at appropriate expectations, and help their trainees make the transition to an appropriate language-learning classroom environment.

2. Consider affective elements: Trainers must be made aware of the importance of understanding the national preferences in terms of learning style. Then, they can help reduce trainees' tension and anxiety and mitigate the culture shock caused by their transition to the communicative language learning process.

3. Consider learner variables: All pedagogical models must incorporate an awareness of individual learner variables—learning styles, multiple intelligences, and left- and right-brain orientation—to give all learners the opportunity to obtain maximum benefit from the instructional design. Trainers must be coached on the implications of each of the learner preferences; they will need to adapt classroom materials so that each trainee's preferred style is used at least some of the time.

4. Employ metacognitive strategies: Instructors should help trainees develop metacognitive awareness of the course structure and the rationale behind it. Thoughtful introspection and the development of a strong metacognitive

knowledge base will help trainees more easily transition from their traditional pedagogical model to the model that is preferred for language training.

5. Bridge the old and new paradigms: It is important for the trainer to find ways to work within the traditional cultural framework of the trainees, especially at the beginning, and then find ways to build bridges to the new pedagogical paradigm so that trainees will be more comfortable as they become more fully engaged in the language-learning process.

6. Transition to a communicative classroom environment: As linguistic research clearly demonstrates, ESP learning is most successful when trainers immerse trainees in a model in which there is interactivity and ample opportunity for the use of the target language in a realistic setting. The final step in implementing this model is to establish a communicative classroom in which language learning can be maximized.

From the cultural perspective as well as the language-teaching perspective, several specialized issues emerge with regard to a blended learning model. The first guideline of the acculturated blended learning model is to train the trainers, but this in itself is a complex requirement. In the case study discussed here, there is a professional need for trainers who have some aviation expertise. In any kind of ESP training, content expertise is at the top of any list of qualifications. This expertise must be strongly backed by knowledge of language training and dedication in providing, in this case for adult trainees (see Kanuka, 2006, for an overview). In the English for special purposes case, the trainers must have a solid background in teaching English as a second or foreign language (TESL or TEFL). While there is a fair amount of research on language-teaching qualifications, Kanuka points out the lack of research focusing on trainers who work within an e-learning environment. As she points out, "Issues of tutor knowledge and teaching methodology are important here . . . : they are required to demonstrate a commitment to communicative language-teaching methodology and to have a broad understanding of the societies and cultures of the country in question" (p. 3). In the site-based section of the blended learning program, trainers must play many roles—not just be the sage on the stage but also the guide on the side—a true facilitator. However, this requires a real transition within any cultural pedagogical framework. Instructors must be taught to anticipate cultural differences, avoid common misunderstandings and miscues, and make short-term adaptations.

With regard to the aviation English case study discussed in this chapter, while the online training program has been successful, there are limited results to report for the blended learning model. As has been evident for the last several years, the aviation industry has suffered severe economic challenges, and these are getting worse with the rising cost of fuel. In most corporations, when the budget gets tight, corporate training is one of the first line items that is reduced or cut out completely. That has certainly been true in the aviation industry, even with English proficiency requirements currently mandated by the International Civil Aviation Organization (ICAO). As a result, the majority of clients for the Aviation English, Inc., product have opted for the online delivery model rather than the blended learning model.

However, a group of Chinese military pilots came to Florida to participate in the blended learning model of aviation English training at the Pan American Flight Academy, which offers the Aviation English, Inc., product. For this blended learning program, the guidelines for acculturated blended learning were implemented. The cultural adaptation was limited slightly limited by the fact that the delivery was in the United States rather than in China, but the ESL instructors were carefully trained in both content delivery and Chinese culture before the group arrived. They implemented metacognitive strategies with the trainees to help them transition to the interactive language-training program. A study analyzing the success of this program documented that the participants achieved significant improvements in their total posttest scores as well as significant increases in each section—listening, grammar, vocabulary, and reading—of the final exam (Strother et al., 2005).

CONCLUSION

Course design using any form of cybereducation—from computer-aided instruction to strictly online delivery to the use of an acculturated blended learning model—must consider a complex collection of elements and the ways in which they interact. Any successful design must, of course, apply the pedagogy appropriate to the course and must be compatible with the cultural dimensions of each group of learners.

Kirkwood and Price (2006) conclude that "technology must provide the tools rather than the drivers for achieving core educational outcomes. If technologies are to be used purposefully to enhance student learning, they need to be integrated not just in terms of pedagogical tactics, but must also reflect and align with the fundamental educational philosophy and aims." However, in Kirkwood and Price's excellent analysis, as is often the case, there is no consideration of the cultural element as it concerns the individual learners, the group, or the cultural context in which the training is to be delivered.

We know that it is impossible to design one online training product that will be ideal for all learners or groups of learners. However, wherever possible, we should attempt to modify our course design to adapt it to cultural learning styles and learner variables wherever possible. For example, if we know we are going to design an online course for East Asian learners, we should first identify the pedagogical patterns preferred in that part of the world and the learner preferences that tend to be characteristic in East Asian cultures.

Using the Felder-Silverman learning style model (1988), we can place learners in the following categories:

- *Sensing*: They are concrete, practical, oriented toward facts and procedures.
- *Intuitive*: They are conceptual, innovative, and oriented toward theories and meanings; they prefer to discover possibilities and relationships; they like innovation and dislike repetition.
- *Visual*: They prefer visual representations of material such as pictures, diagrams, and flow charts.
- *Inductive*: They prefer going from the general to the specific.
- *Active*: They learn by trying things out and working with others.
- *Reflective*: They learn by thinking things through; they work better alone.
- *Sequential*: They learn in linear, orderly, incremental steps.
- *Global*: They learn by thinking globally and holistically, and they learn in large leaps.

How do most Asian students fit into the Felder-Silverman categories? East Asian learning styles tend to be more *sensing,* that is, oriented toward facts and procedures, than intuitive. They tend to prefer *visual* rather than verbal approaches. Asian learners, like those in many communal societies, seem to prefer *deductive* learning styles. They are familiar with classroom approaches in which academic materials are presented in a *sequential* format, which moves from the general rule to specific examples. Asian learners tend to be *reflective:* that is, instead of experimenting or actively trying things out, they tend to learn by thinking things through (see Strother, 2003b, for an overview of this research).

With this knowledge about East Asian learners, how would we adapt an online course for them? To address the preference for the sensing and deductive learning styles, we should design activities to ask concrete factual questions in a variety of question types, with feedback provided for each question giving specific, concrete answers and explanations. For visual learners, online courses are ideal, especially if they incorporate a wide variety of visual elements instead of just text-based pages. Reflective learners can work at their own pace and therefore think through the material carefully. Sequential learners need to have explicit information on the logical flow of individual activities and the overall order of course materials.

The Felder-Silverman learning style model can be used to analyze any cultural group for whom we are developing online training. While individual exceptions are to be found in any country or cultural group, generalizations such as these provide guidance to program designers as they attempt to adapt cybereducation to cultural expectations. Learners are most successful when the instructional methods match individual learning preferences. Learners enjoy the learning experience more, are less frustrated, and stay more motivated—all of which leads to a better learning outcome.

More research should be directed toward determining how cybereducation is affected by cultural factors. This includes everything from Web design to the virtual environment in which students and instructors can or do communicate (see, e.g., Macfadyen, 2005). While cultural differences among learners have not always been taken into account, there is increasing evidence that taking into account these differences constitute a key element in making learning successful for multinational learners.

This particular case study has demonstrated that, along with all the other cultural dimensions that come into play when designing any form of cyber-education, the multiple elements affecting any group of trainees have to be considered, since the members of the group bring overlapping and sometimes contradictory cultural values with them to any training setting, whether online or on site.

In conclusion, I want to reiterate that it is extremely important to consider all facets of culture and the relevant subcultures—national, professional, and pedagogical—as well as individual learner characteristics when designing a Web-based training program. These considerations are critical during all phases of the project. Technical communicators, content developers, and program developers must consider culture as an important part of their audience analysis. Wherever possible, cultural elements—especially those that affect meaning or representations of meaning—must be considered when designing the user interface. In any type of Web-based training, all of the usual pedagogical factors must form part of the curriculum design. In addition, whenever an online program is accompanied by site-based elements, as in a blended learning course with classroom sessions, the cultural preference for pedagogical style should be taken into account. The on-site trainers should prepare and deliver an acculturated blended learning program whenever possible.

If we include all the above facets of culture in the development of the curriculum, the user interface design, and the course implementation, all international/multicultural trainees will have a better chance of success, whatever the training content area.

REFERENCE

Banham, V., & Wong. L. (Eds.). (2001). Blurring borders, respecting culture. In *AARE 2001 International Education Research Conference*. Perth, Australia. Retrieved January 18, 2010, from http://www.aare.edu.au/01pap/won01036.htm

Bourges-Waldegg, P., & Scrivener, S. A. R. (1998). Meaning, the central issue in cross-cultural HCI design. *Interacting with Computers, 9,* 287–309.

Bourges-Waldegg, P., & Scrivener, S.A.R. (2000). Applying and testing an approach to design for culturally diverse user groups. *Interacting with Computers, 13*(2), 111–126.

Buzzetto-More, N., & Sweat-Guy, R. (2006). Hybrid learning defined [Electronic version]. *Journal of Information Technology Education, 5,* 153–156.

Chowdhury, R. (2003). International TESOL training and EFL contexts: The cultural disillusionment factor. *Australian Journal of Education, 47*(3), 283–302.

Cossell-Rice, J., Coleman, M., Shrader, V., Hall, J., Gibb, S., & McBride, R. (2001). Developing Web-based training for a global corporate community. In B. Khan (Ed.), *Web-based training* (pp. 191–202). Englewood Cliffs, NJ: Educational Technology Publications.

Duffy, T. M., & Osman, G. (2005). Designing cross-cultural distance education programs for facilitating pedagogical change: Challenges and strategies. *21st Annual Conference on Distance Teaching and Learning.* Madison, WI. Retrieved November 30, 2008, from http://www.uwex.edu/disted/conference/Resource_library/proceedings/05_1831.pdf

Dziuban, C., Hartman, J., & Moskal, P. (2004) Blended learning [Electronic version]. *Educause Center for Applied Research, Research Bulletin, 7,* 2–12.

Felder, R., & Silverman, L. K. (1988). Learning styles and teaching styles in engineering education. *Engineering Education, 78*(7), 674–681.

Goodfellow, R., & Hewling, A. (2005). Reconceptualising culture in virtual learning environments: From an "essentialist" to a "negotiated" perspective, *E-Learning, 2,* 355–367.

Gudykunst, W. B., Yang, S. M., & Nishida, T. (1987). Culture differences in self-consciousness and self-monitoring. *Communication Research, 14,* 7–36.

Hall, E. T. (1976). *Beyond culture.* New York: Doubleday.

Hall, E. T., & Hall, M. R. (1990). *Understanding cultural differences.* Yarmouth, ME: Intercultural Press.

Heinze, A., & Procter, C. (2004). Reflections on the use of blended learning. *Education in a Changing Environment.* Salford, England. Retrieved January 18, 2010, from http://www.ece.salford.ac.uk/proceedings/papers/ah-04.rtf

Helmreich, R. (2000a). Culture and error in space: Implications from analog environments. *Aviation, Space, and Environmental Medicine, 71*(9), 133–139.

Helmreich, R. (2000b). Culture, threat, and error: Assessing system safety. In *Safety in aviation: The management commitment: Proceedings of a conference.* London: Royal Aeronautical Society.

Helmreich, R., & Merritt, A. (1998). *Culture at work in aviation and medicine: National, organizational, and professional influences.* Brookfield, VT: Ashgate.

Hofstede, G. (1980). *Culture's consequences: International differences in work-related values.* Beverly Hills, CA: Sage.

Hofstede, G. (1991). *Cultures and organizations: Software of the mind.* New York: McGraw-Hill.

Hofstede, G. (2001). *Culture's consequences: Comparing values, behaviors, institutions and organizations across nations* (2nd ed.). London: Sage.

Hofstede, G. (2002). Culture's recent consequences: Using dimension scores in theory and research. *International Journal of Cross-Cultural Management, 1*(1), 11–17. Retrieved January 18, 2010, from http://www.cic.sfu.ca/forum/GeertHofstedeMar082002.html

Hofstede, G., & Hofstede, G. J. (2005). *Cultures and organizations: Software for the mind* (2nd ed.). New York: McGraw-Hill.

Hofstede, G., & McCrae, R. R. (2004). Personality and culture revisited: Linking traits and dimensions of culture. *Cross-Cultural Research, 38*(1), 52–88.

House, R. J., Hanges, P. J., Javidan, M., Dorfman, P. W., & Gupta, V. (Eds.). (2004). *Culture, leadership, and organizations: The GLOBE study of 62 societies.* Thousand Oaks, CA: Sage.

Huang, R., & Leung, F. K. S. (2005). Deconstructing teacher-centeredness and student-centeredness dichotomy: A case study of a Shanghai mathematics lesson. *Mathematics Educator, 15*(2), 35–41.

Kanuka, H. (2006). Issues, challenges and possibilities for academics and tutors at open and distance learning environments. *International Review of Research in Open and Distance Learning, 7*(2). Retrieved September 2006, from http://www.irrodl.org/index.php/irrodl/article/viewArticle/368/642.htm

Kirkwood, A., & Price, L. (2006) Adaptation for a changing environment: Developing learning and teaching with information and communication technologies, *International Review of Research in Open and Distance Learning 7*(2). Retrieved January 18, 2010, from http://oro.open.ac.uk/6216/01/Adaptation_for_a_changing_environment PDF.pdf

Borytko, N. (1991). Values and education: Russian perspective. *Zbornik Instituta za Pedagoška Istraživanj, 37,* 35–56.

Kukulska-Hulme, A. (2000). Communication with users: Insights from second language acquisition. *Interacting with Computers, 12*(6), 587–599.

Lanham, E., & Zhou, W. (2003). Blended learning for cross-cultural e-learning. In G. Richards (Ed.), *Proceedings of World Conference on E-Learning in Corporate, Government, Healthcare, and Higher Education* (pp. 1927–1930). Chesapeake, VA: Association for the Advancement of Computing in Education.

LISA (Localization Industry Standards Association). (2002). LISA homepage. Retrieved May 10, 2008, from http://lisa.org/leit/terminology.html

Macfadyen, L. P. (2005). Internet-mediated communication at the cultural interface [Electronic version]. In C. Ghaoui (Ed.), *Encyclopedia of human-computer interaction* (pp. 373–380). Hershey, PA: IGI Global.

McCarty, S. (2005, April). Cultural, disciplinary and temporal contexts of e-learning and English as a foreign language. *eLearn Magazine, 4*. Retrieved January 18, 2010, from http://portal.acm.org/citation.cfm?id=1070948.1070950

Merritt, A. (2000). Culture in the cockpit: Do Hofstede's dimensions replicate? *Journal of Cross-Cultural Psychology, 3,* 283–301.

Morse, K. (2003). Does one size fit all? Exploring asynchronous learning in a multicultural environment. *Journal for Asynchronous Learning Networks, 7*(1), 37–55.

Niles, S. (2005). Cultural differences in learning motivations and learning strategies: A comparison of overseas and Australian students at an Australian university. *International Journal of Intercultural Relations, 19*(3), 369–385.

Onibere, E. A., Morgan, S., Busang, E. M., & Mpoeleng, D. (2001). Human-computer interface design issues for a multi-cultural and multi-lingual English speaking country—Botswana. *Interacting with Computers, 13*(4), 497–512.

Payne, C. (2002). *The e-learning market in Asia-Pacific.* Retrieved May 10, 2008, from http://www.apconnections.com

Planel, C. (1997). National cultural values and their role in learning: A comparative ethnographic study of state primary schooling in England and France. *Comparative Education, 33*(3), 349–373.

Ramburuth, P. (2000). *Cross cultural learning behaviour in higher education: Perceptions versus practice.* Lecture presented to the Seventh International Literacy and Education Research Network Conference on Learning, RMIT University, Melbourne,

Australia. Retrieved January 18, 2010, from http://ultibase.rmit.edu.au/Articles/may01/ramburuth1.htm

Rosman, A., & Rubel, P. G. (1995). *The tapestry of culture: An introduction to cultural anthropology* (5th ed.). New York: McGraw-Hill.

Rossett, A., & Franzcc, R. V. (2006). Blended learning opportunities. American/Management Association. Retrieved January 18, 2010, from http://www.amanet.org/training/seminars/Blended_Learning_Opporunities-45.aspx

Rovani, A., & Jordan, H. (2004). Blended learning and sense of community: A comparative analysis with traditional and fully online graduate courses. *International Review of Research in Open and Distance Learning*. Retrieved from http://madlib.athabascau.ca/oldirrodl/content/v5.2/rovai-jordan.html

Schwartz, S. H. (2004). Beyond individualism/collectivism: New cultural dimensions of values. In U. Kim, H. C. Triandix, C. Kagitçibasi, S. C. Choi & G. Yoon, (Eds.), *Individualism and collectivism Theory, method and applications*. Thousand Oaks, CA: Sage.

Schwartz, S. H., & Bardi, A. (2001). Value hierarchies across cultures: Taking a similarities perspective. *Journal of Cross-Cultural Psychology, 32*(3), 268–290.

Shaw, W., & Igneri, N. (2005). *Effectively implementing a blended learning approach: Maximizing advantages and eliminating disadvantages*. Eedo Knowledgeware and American Management Association. Retrieved from http://adlcommunity.net/file.php/11/Documents/Eedo_Knowledgeware_whitepaper_Blended_Learning_ AMA.pdf

Sherman, P., Helmreich, R., & Merritt, A. (1997). National culture and flight deck automation: Results of a multination survey. *International Journal of Aviation Psychology, 7*(4), 311–329.

St.Amant, K. (2005, January). A prototype theory approach to international website analysis and design. *Technical Communication Quarterly, 14*(1), 73-91.

Strother, J. B. (2003a). Cross-cultural issues for Asian e-learners: An analysis based on Hofstede's cultural dimensions. *The e-Learn: AACE World Conference on E-Learning in Corporate, Government, Healthcare, and Higher Education* (pp. 1978–1984). Phoenix, AZ. Retrieved January 18, 2010, from http://ieeexplore.ieee.org/Xplore/login.jsp?url=http%3A%2F%2Fieeexplore.ieee.org%2Fiel5%2F8817%2F27908%2F01245513.pdf%3Farnumber%3D1245513&authDecision=-203

Strother, J. B. (2003b). Shaping blended learning pedagogy for East Asian learning styles. *2003 IEEE International Professional Communication Conference* (pp. 353–357). Orlando, FL. Retrieved January 18, 2010, from http://ieeexplore.ieee.org/xpl/freeabs_all.jsp?arnumber=1245513

Strother, J. B., & Alford, R. L. (2003). Addressing learner variables in an e-learning environment. *The e-Learn: AACE World Conference on E-Learning in Corporate, Government, Healthcare, and Higher Education* (pp. 1971–1977). Phoenix, AZ. Retrieved January 18, 2010, from http://my.fit.edu/~strother/downloads/Addressing_Learner_Variables.pdf

Strother, J. B., Chiu, C., Serrano, C., & Alford, R. (2005). Effectiveness of implementing computer-assisted language learning technology in an English for specific purposes training program. *International Journal of Learning, 11*. (Originally presented at the 11th International Literacy and Education Research Network Conference on Learning in Havana, Cuba, July 2004.)

Strother, J. B., Fazal, Z., & Gurevich, M. (2007). Acculturated blended learning: Localizing a blended learning course for Russian trainees. The *IASTED Conference.* Chamonix, France. Retrieved January 18, 2010, from http://portal.acm.org/citation.cfm?id=1323240

Tabulawa, R. (1998). Teachers' perspectives on classroom practice in Botswana: Implication for pedagogical change. *International Journal of Qualitative Studies in Education, 11*(2), 249–268.

Tabulawa, R. (2003). International aid agencies, learner-centred pedagogy and political democratization: A critique. *Comparative Education, 39*(1), 7–26.

Triandis, H. C. (2004). The many dimensions of culture. *Academy of Management Executive, 18*(1), 88–93.

Triandis, H. C., & Tratimow, D. (2001). Cross-national prevalence of collectivism. In C. Sedikides & M. B. Brewer (Eds.), *Individual self, relational self, collective self* (pp. 259–276). Philadelphia: Psychology Press.

Volet, S., & Renshaw, P. (1996). Chinese students at an Australian university: Adaptability and continuity. In D. A. Watkins & J. B. Biggs (Eds.), *The Chinese learner: Cultural, psychological and contextual influences* (pp. 205–220). Hong Kong, China: CERC and ACER.

Yamada, A., & Singelis, T. (1999). Biculturalism and self-construal. *International Journal of Intercultural Relations, 23*, 697–709.

Yong, K. (2003). *Culture issues in Taiwan's civil aviation community*. Retrieved from www.itsasafety.org/data/2003/Culture_study_in_Taiwan.doc

Zhao, Y. & Seppo, T. (2002). From the special issue editors. *Language, Learning and Technology, 6,* 1–4.

CHAPTER 10

Digital Ecologies: Observations of Intercultural Interactions in Learning Management Systems

Sipai Klein and Sharon Trujillo Lalla

CHAPTER OVERVIEW

Learning management systems are becoming widely adopted across the globe, and as educational institutions adopt these technologies, the question arises as to whether these technologies meet the needs of globally diverse workplaces and classrooms. Globalization implies a global platform, a network approach that encourages the endorsement of diverse values, social practices, and communication modes. This chapter uses intercultural variables from Hall (1976), Hofstede (1980), and Hampden-Turner and Trompenaars (2000) to provide an intercultural density of learning management systems. The authors of this chapter provide observations for technical communicators and educators who facilitate and design training and curriculum for people from diverse cultures in online learning environments. It is hoped that this description of the affordances and disadvantages of learning management systems for intercultural online interactions will make the need to bridge the gap in intercultural communication clear to future researchers.

Used throughout the globe by companies and educational institutions, learning management system technologies are prime sites for a discussion about intercultural communication. How intercultural variables influence usability and receptibility and how intercultural interaction occurs in these systems are the issues and scenarios addressed in this chapter. An understanding of the intercultural considerations encountered by a user of a learning management system (LMS) may assist technical communicators and educators in facing the challenges. Whether an LMS is part of an online training workshop or of a hybrid course that combines both face-to-face and online environments, this chapter aims to provide an intercultural density, namely, a degree of both opacity and transparency, of cultural interaction in learning management systems.

Just as important as the wide adoption of an LMS are the recent technological advances that allow such systems to serve as multimedia authoring tools where technical communicators and educators may make choices on how to situate content and communicate with participants. For example, instructors may redesign icons, employ blogs to generate discussion with images, create hyperlinks to connect internal content to other content, and designate an order for online activities. The capacity for authoring decisions empowers people to make user-based choices. Especially since online scenarios tend to magnify "culturally based style in intercultural interactions" (Pekerti & Thomas, 2003, p. 150), the new technologies afford opportunities for people to make clear decisions on how to interact with an audience and how to design interactive activities and establish communicative expectations among participants from diverse cultures.

Our discussion of learning management systems develops from an understanding that cultures do not exist between dichotomous poles but rather are rich with possibilities that may be described as ecologies. The wealth of salient and explicit talents in each culture can be understood when cultures are brought together through interaction, which in this case occurs in a learning management system environment. Just as the chemical properties of oil and water are better understood when they are in combination, so too do cultures enlighten us when people with multiple frames of reference participate and share their ideas in an online environment.

This chapter considers various cultural variables, such as individualism, collectivism, context, presence, and diffusiveness—all of which are prevalent in online environments. We begin by discussing how intercultural communication evolved from cross-cultural studies, in which single cultural frames of reference are analyzed without interaction, to the point at which cultures are examined as interacting frames of reference (Hewling, 2006). We then provide a series of observations on the affordances and disadvantages of learning management systems when users are culturally diverse and offer examples that can be applied in online environments. The chapter concludes with a real scenario showing how intercultural considerations may assist technical communicators and educators as they develop curriculum and training for culturally diverse users.

A DISCUSSION OF ONLINE INTERCULTURAL INTERACTION

Our discussion of online intercultural interaction in learning management system technologies subscribes to Hall's (1976) view that culture is a learned, interrelated, and shared experience. In turn, culture is the medium with which we express ourselves in person or in an e-mail; it is the means by which we solve problems by ourselves or as members of a listserv and by which we organize information in a handbag or in a content module. The differences that designate group boundaries allow us and other researchers to study cultures and make culture-specific choices. Like St.Amant (2002), who relies on theory and research by Hall and his successor Hofstede (1980), we also rely on concepts and terminology developed by previous researchers, yet we develop our ideas by using a situational model in which two or more groups are brought to interact together.

The history of intercultural research, namely, the study of the interaction of cultures, is founded on Hofstede's (1980) 20-year research project at IBM, where he analyzed employees from diverse cultures. Hofstede employed quantitatively coded data to analyze participants' responses to similar hypothetical scenarios. Visually, then, cultures are illustrated as normalized distributional curves with each curve representing a trend. The basic assumption of such a representation is that when two cultures are illustrated as curves with amplitudes and wavelengths, a researcher may pinpoint two polarities and describe two groups of participants as cultures existing between dichotomous poles. This assumption, exemplified in Hofstede (1980) and his successors (e.g., Hampden-Turner & Trompenaars, 2000), help to systematically examine a variety of cultures, yet criticism has been expressed with regard to the limitations of polarizing cultures; some intercultural communicators and rhetoricians have taken these limitations into consideration when devising their own research. Ratner (2006), for example, critiques the construction of culture as an amalgamation of singular variables and elaborates on the cat-and-mouse game between those who believe that cultures develop from micro, psychological levels and those who believe cultures to be immutable contexts reified by individual behavior.

Faiola and Matei (2005) argue that Hofstede's five cultural dimensions are derived from behavioral psychology's narrow definition of culture by correlating culture with geography. They rely on cognitive psychologists Nisbett and Norenzayan (2002), who have developed the idea of cultural psychology to explain social practices as derived from Vygotsky's (1978) theory of learning as a shared process. Faiola and Matei define cultural differences as preferences arrived at with regard to material-based classification or shape-based classification, field-dependence or field-independence styles, and holistic or serialist problem-solving approaches. Ultimately, the authors state, "cultural practices and cognitive processes constitute one another" (2005, p. 13) and, later on, they even correlate cultural cognitive theory with Web design (Faiola & Matei, 2006): "Besides the explicit cultural differences of text, numbers, dates, symbol-sets, and time, more critical are the implicit and less formal dimensions of page format, imagery, color, information architecture, and system interaction" (2006, p. 15). In a similar vein, also straying away from binary cultural variables, other researchers state that "cultural analyses resting on such relatively simple dichotomies may be too simple for dealing with the real-world complexities of culture" (Ess & Sudweeks, 2006, p. 180).

INDIVIDUALISM AND COLLECTIVISM

The cultural elements of individualism and collectivism can be used as lenses through which to view the features of learning management systems. According to Hermeking (2005), Internet usage has a strong correlation with individualistic cultures, whose features are commonly employed in online learning environments. The self-reliant, competitive, expressive features of individualism can be found when a student submits an assignment on his or her own or keeps a journal on a private online space. If a grade-posting feature is included in a learning management system and the grade shows only an individual student's standing, then the message is that it is the student's individual self-development, and not the group interaction, that is evaluated. In addition, learning management systems generally delegate a personal tone to many of their learning features, such as My Grades and My Progress, that emphasize the individual's assessment. Probably the most salient feature of individualism in this online learning environment is the detachment from a sense of social interaction conveyed by the interface's design. When logging on to a class, a student sees a home page instead of other students or the class instructor. Each student logs on to the class on his or her own time, at his or her computer, and, in an online-only class, at a location of his or her choice.

Students and facilitators who practice in collective cultures expect features involving cooperative, interpersonal, and societal legacy values. Elements of collectivism in a learning management system can present themselves through online, synchronous activity instead of through the use of discrete features. The immediate presence of others, who appear in chat rooms and through instant messaging, is available with some learning management systems. A student or facilitator need only to coordinate a time with other students in order to communicate in this manner within a learning management system. Facilitators can build in virtual office hours, virtual white boards, and virtual lectures to bring a collective audience together to share and exchange ideas. Successful synchronous activities, however, involve strategies for engaging students online.

Asynchronous activities may also support collective efforts, yet Warden, Chen, and Caskey (2005) and Shih and Cifuentes (2003) have pointed out that discussion boards do not support the needs of collective cultures. They found that Asians are at a disadvantage within discussion forums due to their communication norms. Ku and Lohr's (2003) research on Chinese students using a learning management system to interact with American students also identifies the isolation experienced online. Consider, however, that discussion boards can significantly contribute to collective goals if sound, well thought-out discussion strategies are set in place to encourage and motivate students to participate in group work. As a matter of sound instructional design, motivational requirements such as clear accountability and participation guidelines must be tied to assessment and sewn into the fabric of the online course with regular follow-up by the facilitator. In intercultural situations, students should become aware of the thinking patterns, expressions, and characteristics of the members of other cultures who are involved (Chen, 1998; St.Amant, 2002). Integrating intercultural awareness into the course assignments as a natural expectation of the course can help to alleviate misunderstandings.

Members of individualistic cultures tend to respond to agency-provoking images and messages, while those from collective cultures tend to respond to community-provoking design (Cook & Finlayson, 2005). Building a successful online community, therefore, requires close attention from the designers of an online course. Facilitators and technical communicators might consider designing the online course with greater concern for societal legacy, for example, adding cultural images to banners and background images, adding national or heritage-related commemorative days to calendar announcements, and addressing personal intentions in announcements posted on the homepage.

HIGH-CONTEXT AND LOW-CONTEXT COMMUNICATION

Differences between cultures appear when we consider Hall's (1976) context theory: "A high-context (HC) communication or message is one in which most of the information is either in the physical context or internalized in the person while very little is in the coded, explicit, transmitted part of the message. A low-context (LC) communication is just the opposite; that is, the mass of the information is vested in the explicit code" (p. 91). An LMS is not self-explanatory, though it does contain instructional information on most command buttons, which explicitly direct the user to compose, compile, attach, and so on. The buttons communicate explicit functions and therefore help produce a low-context learning environment. The knowledge base required for the user to navigate through the program is limited to these explicit functions. This information is decontextualized, and the directives assume that the user understands what it is that will be composed, compiled, or attached, for example. Aside from the interface, examples of low-context communication instances in an LMS include course syllabi, calendars of events, general announcements, supplementary readings, and grade books.

A high-context communication, on the other hand, is vested in the meaning of contextualized messages, in which implications and inferences result in a variety of individualized meanings. For example, failing to respond to a discussion message when the expectation is a reply may represent exclusion for the person who sent the original message. The high-context communication devices in an LMS include tools involving interaction between the student, the teacher, and the content. Examples include e-mail, online discussion boards, synchronous text chats, and content modules—environments where students interact with the content in a course. An online discussion, for example, revolving around a particular topic can result in highly contextualized messages. Users evoke contexts when providing constructive criticism or offering compliments. In some cultures, it is more important to *save face* than to expose one's weaknesses in the guise of constructive critique (Kim, Hearn, Hatcher, & Weber, 1999). Differing perceptions can be significant challenges in intercultural exchanges, as was the case when international students from New York and New Zealand collaborated in online projects. Perceptions of urgency, stress, and required level of analysis became problematic (Zhu, Gareis, Bazzoni, & Rolland, 2005).

Low- and high-context communication opportunities exist in learning environments involving cultural and intercultural differences. In a learning environment provided by a tool such as an LMS, the ease with which the user interface can be used by a broad audience of users should minimize low-context communication problems. This can also be achieved through technological readiness—preparing users to use the technology. User interaction and the diversity of the learners, however, will situate communication in a dynamic, highly contextual learning environment.

POWER MANAGEMENT

Effective intercultural communication is indeed complex and varied. This can be demonstrated by applying Hall's (1976) context theory to a simple communication tool such as e-mail. Kim et al. (1999) found cultural differences while managing power in e-mail communications among Koreans and Australians. The Koreans demonstrated the use of an e-mail protocol that is prevalent in a culture where relationships are most important. Korean e-mail users assumed unequal distribution in institutional and organizational power. They were more cognizant of hierarchical structure, resulting in a formal, respectful, and limited form of communication in their e-mails. The Australian e-mail users, on the other hand, demonstrated the use of an e-mail protocol reflecting low-context communication. Australian e-mail users were generally less formal and direct, believing that the main purpose of the e-mail tool was to convey meaning. In response to the conflict with regard to power, Koreans often chose silence.

Similarly, Thorne's (2003) intercultural study found that e-mail is effective only for some power relationships while ineffective for others. His research found that both French and U.S. students perceive e-mail as an effective communication tool for students communicating with teachers or for sons or daughters communicating with parents. They did not, however, believe that e-mail was suitable for creating intercultural peer relationships or social interactions. Both French and U.S. students preferred alternative computer-mediated devices such as instant messaging to create intercultural relationships.

ONLINE PRESENCE

Though their presence in all of its physical meaning is displaced in an online environment, LMS participants leave digital traces of their activities in the system when they log in and out, hit on links, post messages, send e-mails, take quizzes, and actively engage within the system. At times, participants are limited to textual communication for the purpose of identifying themselves and recognizing others. Merryfield (2001) insightfully wonders if there are "specific, qualitative differences in understanding another person, especially one culture from another" (p. 296) when people interact face-to-face and employ body language as opposed to when they interact through online text messages. Students who interact in an online learning environment with other students from around the globe cannot employ body language to communicate the full spectrum of their presence. When logging on to an LMS, especially one that heavily relies on asynchronous communication modes, users interact with the system mainly through written curricular tasks. Relying only on written text communication to impart presence in an online environment raises the same concern that Hall (1976) raised about U.S. foreign language education that ignored a holistic view of culture and focused instead on piecemeal heuristic approaches.

In newer learning management systems, synchronous communications tools are more readily integrated into the overall system, thus removing concerns over a lack of immediate presence. Designers of courses who enable instant messaging to see who has logged on to the system allow facilitators to communicate instantly with participants and also allow participants to communicate with facilitators. Such integration, while delivering the potential for higher accessibility among the participants, also raises the issue of the type of written language used to communicate in an online environment. Kress (2003) points out that the immediacy of some computer-mediated communication tools, such as e-mail and instant messaging, allows for more informal language. Participants, then, might employ more speech-like text to communicate their presence in an online environment, presenting a new level of intercultural interaction and challenge. A more colloquial language provides a richer means of establishing presence and recognizing the cultural landscape of participants, yet it does not necessarily meet the curricular goals of the facilitators. Another interesting interaction is the integration of multimodal texts, such as the one used in the *cultura* model of telecollaboration (O'Dowd, 2005), where participants supply information such as opinion polls, press articles, and images to their peers. This allows for a more holistic view of establishing presence and reduces the invisibility of participants.

SPECIFIC AND DIFFUSE

Learning management systems contain the tensions traditionally existing between diffuse-oriented and specific-oriented cultures. Hampden-Turner and Trompenaars' (2000) definition of specificity and diffusiveness is that "While specificity focuses on the product, diffusiveness considers the entire process by which the product is conceived, designed, developed, manufactured, distributed, and maintained" (p. 137). Furthermore, specific-oriented cultures employ a direct communication style, while diffuse-oriented cultures use an indirect communication style that relies on the listener to fill in the interpretation. Finally, members of specific-oriented cultures more easily share their private lives with the public, while members of diffuse-oriented cultures retain a much larger private space that is closed off to the public. These tensions can be seen in both the interface and the devices in learning management systems.

Like diffuse-oriented cultures, LMS technology attempts to network among its various functions and focus on the process by which learning is shared and created. For example, a basic assumption of the discussion board tools found in an LMS is the integration of ideas by participants. Similarly, designers may interlink various functions into modules in which users are expected to link reading materials, evaluative tasks (such as quizzes), and communication activities. This sort of integrated approach to LMS design emphasizes the "system" part of the technology and reveals characteristics of diffuse-oriented cultures.

Yet, characteristics of specific-oriented cultures may also be found in an LMS. As previously noted, the interface describes itself in terms of its functionality—its potential as a product-making device. More than a rapport with other users, the interface gathers information on participants' activities with the aim of composing a report. How many messages were posted, how many tasks met a facilitator-designated rubric, and how participants scored on each activity tends to objectify users and avoid the indirect language that hopes others will fill in the correct interpretation. Some learning management systems even assign a grade per discussion board posting, thus emphasizing users' capacity to deliver products as opposed to their capacity to develop relationships. All of these are characteristics of specific-oriented cultures.

Another reflection of the tension between diffuse-oriented and specific-oriented cultural characteristics in an LMS is the use and establishment of private and public spaces. Much of the space available in an LMS is monitored. For example, messages on discussion boards can be viewed by all; this is a characteristic of specific-oriented cultures, in which personal space tends to be brought into public view. At the same time, most of the student's private life is invisible to others. Unlike users of online social networks such as Myspace and Facebook, LMS participants do not have space to designate as private and by default to segment their private lives. Such a distinction between an individual's private and public space, while designers and facilitators possess the capacity to designate spaces where thoughts and feelings may be expressed in public, reflects the tension between specific-oriented and diffuse-oriented cultural characteristics that may occur in an LMS.

VISUAL LANGUAGE

Visual language in online environments directly addresses the need for cultural sensitivity (Cook & Finlayson, 2005; Marcus, 2003; Marcus & Baumgartner, 2004; Marcus & Gould, 2000; St.Amant, 2005), especially in terms of design (Bourges-Waldegg & Scrivener, 1998; Chen, Mashhadi, Ang, & Harkrider, 1999; Collis, 1999). Research by Faiola and Matei (2005, 2006) indicates that cultures differ in terms of how they read information on the screen: "In addition to the explicit cultural differences of text, numbers, dates, symbol-sets, and time, more critical are the implicit and less formal dimensions of page format, imagery, color, information architecture and system interaction" (2006, p. 380). Not only are the types of visual choices important, but also the numbers of images and written texts need to be considered, as shown by Onibere, Morgan, Busang, and Mpoeleng (2001), whose research indicates that some cultures are "more disposed to the use of menus and language-based interfaces" (p. 509). Designers need to consider, for example, that *white* symbolizes purity in the United States, while in Japan it represents death (Cook & Finlayson, 2005), or that the color *red* is a symbol of good luck in Chinese culture (Dragga, 1999). Icons, wallpapers, fonts, layout, color patterns, and information hierarchies may all serve cultural meaning-making functions. Educators and technical communicators are, at times, limited to the options provided by the learning management system, though usually they do have options. They may choose from a database of images in order to redesign icons and background colors. Ultimately, visual language choices allow designers to represent meaning in specific contexts, following the advice of Bourges-Waldegg and Scrivener (1998), whose research shows that "culturally determined usability problems are centered in how the representations are used in a system" (p. 290). That is, educators and technical communicators who consider how visual language represents meaning in a specific context are more likely to successfully address the needs of culturally diverse audiences.

INTERCULTURAL INTERACTION IN A PROJECT-BASED ONLINE ASSIGNMENT

At the very least, a basic understanding of intercultural considerations may assist technical communicators and educators as they develop training and curriculum for culturally diverse users of learning management systems. This section gives an example of an online, problem-based, inquiry assignment that addresses some of the intercultural considerations mentioned in this chapter. We assume that this is the first major assignment in the course.

A problem-based assignment is one reflecting real life and real problems through inquiry and critical analysis. It generally involves a group project and small groups of students. Students are asked to engage in problem-oriented projects to discuss, analyze, and form conclusions. Perhaps this is an oversimplification, but we are dividing the discussion of this type of online assignment into four parts: (1) defining the learning outcomes by setting up the goal(s) and the objectives required for the project; (2) defining the assessment criteria; (3) creating learning activities; and (4) designing content.

Learning Outcomes

As in traditional classroom assignments, we begin the online assignment with stated learning outcomes. Learning management systems provide learning outcome features that are invisible in a face-to-face classroom but that can address the intercultural needs of members of low context, specific-oriented, and individualistic cultures by initially looking at objects before focusing on relationships. In a learning management system, learning outcomes can often be defined through learning outcome tools such as the goal and objective tools. By utilizing a wizard—a user interface consisting of user-led dialog boxes prompting for content, we can readily address the goal(s) and objectives of the course during the planning stage. Although learning outcomes are a basic instructional design component, setting learning outcomes is made easier and more visible through learning management systems. As a result, members of low-context, individualistic, and specific-oriented cultures can begin their inquiry about the end product, the submitted assignment, during the initial phase of the project.

Assessment

In the *early* stages of this assignment, the assessment strategy is extremely important for members of low-context, specific-oriented, and individualistic cultures. Specifically to meet this need, we provide a clearly defined rubric explaining how this assignment will be evaluated, and we make it available at the same time that the assignment is given. In addition, we measure learning through a loosely defined reflective component that is required from each group member after submission of the group assignment. This component allows members of specific-oriented, individualistic, and diffuse-oriented cultures to make interpretations based on personal judgment. Because the problem-based assignment requires a significant amount of group work, we also embed opportunities for collective and diffuse-oriented cultures to begin establishing relationships early on. In the rubric, the points associated with the group interaction are weighted equally.

Learning Activities

In this online problem-based group assignment, discussion and a follow-up report are core activities, with students required to interact substantially to complete this project. In an LMS, discussion boards are designed to generate interactivity and as a means of posting content in, albeit restricted, public and private forums. Simply by incorporating both private and public spaces for group interaction, we can offer flexibility to users from high-context, collective, diffuse-oriented, and individualistic cultures.

We begin the problem-based assignment with group activity. Members of collective and diffuse-oriented cultures are made more at ease at the beginning of the course with group-building activities and interpersonal introductions. In a public forum, students are asked to present an artifact from their culture representing successful projects from their hometown or country of origin. In such forums, students increase their awareness of how various cultures view success and accomplishments and are encouraged to continue discussing and building relationships. In addition, individuals can express themselves, so as to create a sense of personal agency.

Standard features in discussion forums also include the ability to create private spaces. In this assignment, students self-select group membership based on the problem topic that interests them. We then provide each team with a private virtual space. A team-building communication exercise geared to collective and diffuse-oriented cultures, where relationships are crucial, is given, so that students continue to learn more about each other. This exercise focuses solely on student-to-student interaction as students learn how other students prefer to work in a team. We also require that group members meet synchronously in chat rooms, where a transcript is made of their conversations and colloquial language is more engaging, to demonstrate their contributions to the team. Users from specific-oriented and individualistic cultures might appreciate the chat record requirement to demonstrate their contributions, while users from collective and diffuse-oriented cultures might welcome the review of communication exchanges.

In this group project, we give student groups the flexibility to submit their products using a delivery mode that best suits their needs. Expanding the delivery modes afforded by the technology, we provide students from collective cultures with an assignment that allows teams to share in the decision about the delivery method used, which might include audio, video, and text presentation. This expansion in delivery modes gives students more flexibility in the approach they choose to showcase their knowledge of a particular topic. Incorporating modes of presentation that go beyond text, students can present information in ways that have deeper cultural density. Such flexibility allows for an increased integration of cultural awareness of how messages are communicated in forms other than written information, for example, image and sound. Such media allow high-context communication scenarios to be integrated with the naturally low-context communication design of the LMS interface. The rubric tied to this assignment contextualizes specific-oriented and individualistic cultures during this decision-making process.

Content Design

We recommend that the design of the assignment be kept simple and consistent in an effort to meet the needs of students from diverse cultures. The description of the assignment, together with its goal(s), objectives, rubric, technology resources, and other resources required to complete it should be easily accessible and display a consistent design. The assignment should also be consistent with the design of the content for the entire course. This may reduce the level of cognitive frustration for low-context, specific-oriented students. Where schools and organizations apply standards to their materials, we can take advantage of general content, such as technological instruction, that is already available. Such an approach visually communicates to students that they are affiliated with an institution or organization, thus meeting the needs of students from collective cultures.

CONCLUSION

The verbal content of a message is as influential for human behavior as the environment in which messages are conveyed (Hall, 1976). The social ecology of an LMS is as purpose driven as the architectural design of the office space in a law firm. Such situated considerations influence usability practices for different cultures. Because students cannot redesign their online learning environments, technical communicators and educators need to consider the limitations and opportunities that learning management systems bring to online intercultural interaction.

We find, on the one hand, that the instructional content students experience in learning management systems confronts the cultural languages of students. We also find that the design of learning management systems also sends messages about the pedagogical approaches that are valued in these types of environments. Studies mentioned in this chapter have examined interactivity among students using computer-mediated communication and have found cultural difficulties in the formation of individualistic or collective identities and contexts fitting for the audience. To meet individualistic and specific requirements, numerous studies mentioned in this chapter reinforce the need to provide concise instruction so as to guide self-directed users to learn about the expectations of the online activity and to use required technologies. Social activities that actively engage students in virtual classrooms provide opportunities for collective and diffuse-oriented audiences to successfully contribute as group members. In addition, it is recommended that virtual classrooms recognize the importance of visual language and integrate intercultural discussions in an effort to understand varied cultural perceptions and values.

Finally, learning management systems carry the potential of generating new ecologies if they are seen as opportunities to explore intercommunicative processes from within these globally distributed learning environments. Interactions among students in these environments provide a close look at learning and communication modes in various cultures. These interactions, in turn, serve as learning experiences, teaching lessons about cultural values and, in and of themselves, about intercultural communication.

REFERENCES

Bourges-Waldegg, P., & Scrivener, S. A. R. (1998). Meaning, the central issue in cross-cultural HCI design. *Interacting with Computers, 9*(3), 287–309.

Chen, A.-Y., Mashhadi, A., Ang, D., & Harkrider, N. (1999). Cultural issues in the design of technology-enhanced learning systems. *British Journal of Educational Technology, 30*(3), 217–230.

Chen, G.-M. (1998). Intercultural communication via e-mail debate. *The Edge: The E-Journal of Intercultural Relations, 1*(4). Retrieved December 1, 2006 from http://www.interculturalrelations.com/v1i4Fall1998/f98chen.htm

Collis, B. (1999). Designing for differences: Cultural issues in the design of WWW-based course-support sites. *British Journal of Educational Technology, 30*(3), 201–215.

Cook, J., & Finlayson, M. (2005). The impact of cultural diversity on Web site design. *SAM Advanced Management Journal, 70*(3), 15–23.

Dragga, S. (1999). Ethical intercultural technical communication: Looking through the lens of Confucian ethics. *Technical Communication Quarterly, 8*(4), 365–381.

Ess, C., & Sudweeks, F. (2006). Culture and computer-mediated communication: Toward new understandings. *Journal of Computer-Mediated Communication, 11*(1), 179–191.

Faiola, A., & Matei, S. A. (2005, May). The cultural cognitive style of multimedia development: Identifying designer cognitive structures for the Web. In K. M. Lee (Chair), *Communication: Questioning the dialogue. Symposium conducted at the International Communication Association* (conference). New York, NY.

Faiola, A., & Matei, S. A. (2006). Cultural cognitive style and Web design: Beyond a behavioral inquiry into computer-mediated communication. *Journal of Computer-Mediated Communication, 11*(1), 375–394.

Hall, E. T. (1976). *Beyond culture*. New York: Doubleday.

Hampden-Turner, C. M., & Trompenaars, F. (2000). *Building cross-cultural competence: How to create wealth from conflicting values.* New Haven, CT: Yale University Press.

Hermeking, M. (2005). Culture and Internet consumption: Contributions from cross-cultural marketing and advertising research [Electronic version]. *Journal of Computer-Mediated Communication, 11*(1).

Hewling, A. (2006). Culture in the online class: Using message analysis to look beyond nationality-based frames of reference. *Journal of Computer-Mediated Communication, 11*(1), 337–356.

Hofstede, G. H. (1980). *Culture's consequences: International differences in work-related values.* Newbury Park, CA: Sage.

Kim, H.-S., Hearn, G., Hatcher, C., & Weber, I. (1999). Online communication between Australians and Koreans: Learning to manage differences that matter. *World Communication, 28*(4), 48–68.

Kress, G. (2003). *Literacy in the new media age.* New York: Routledge.

Ku, H.-Y., & Lohr, L. L. (2003). A case study of Chinese students' attitude toward their first online learning experience. *Educational Technology Research and Development, 51*(3), 95–102.

Marcus, A. (2003). Are you cultured? Global Web design and the dimensions of culture. *New Architect, 8*(3), 28–31.

Marcus, A., & Baumgartner, V. J. (2004). A visible language analysis of user-interface design components and culture dimensions. *Visible Language, 38*(1), 1–65.

Marcus, A., & Gould, E. W. (2000). Crosscurrents: Cultural dimensions and global Web user-interface design. *Interactions, 4,* 32–46.

Merryfield, M. M. (2001). The paradoxes of teaching a multicultural education course online. *Journal of Teacher Education, 52*(4), 283–299.

Nisbett, R. E., & Norenzayan, A. (2002). Culture and cognition. In D. Medin & H. Pashler (Eds.), *Stevens' handbook of experimental psychology: Vol. 4, Memory and cognitive processes* (3rd ed.). New York: John Wiley & Sons.

O'Dowd, R. (2005). Negotiating sociocultural and institutional contexts: The case of Spanish-American telecollaboration. *Language and Intercultural Communication, 5*(1), 40–56.

Onibere, E. A., Morgan, A., Busang, E. M., & Mpoeleng, D. (2001). Human-computer interface design issues for a multi-cultural and multi-lingual English-speaking country—Botswana. *Interacting with Computers, 13*(4), 497–512.

Pekerti, A. A., & Thomas, D. C. (2003). Communication in intercultural interaction: An empirical investigation of idiocentric and sociocentric communication styles. *Journal of Cross-Cultural Psychology, 34*(2), 139–154.

Ratner, C. (2006). *Cultural psychology: A perspective on psychological functioning and social reform.* New Jersey: Lawrence Erlbaum Associates.

Shih, Y.-C. D., & Cifuentes, L. (2003). Taiwanese intercultural phenomena and issues in a United States-Taiwan telecommunication partnership. *Educational Technology Research and Development, 51*(3), 82–89.

St.Amant, K. (2002). Integrating intercultural online learning experiences into the computer classroom. *Technical Communication Quarterly, 11*(3), 289–315.

St.Amant, K. (2005). A prototype theory approach to international Web site analysis and design. *Technical Communication Quarterly, 14*(1), 73–91.

Thorne, S. L. (2003). Artifacts and cultures-of-use in intercultural communication. *Language, Learning and Technology, 7*(2), 38–62.

Trompenaars, F., & Hampden-Turner, C. (1998). *Riding the waves of culture: Understanding cultural diversity in global business* (2nd ed.). New York: McGraw-Hill.

Vygotsky, L. S. (1978). *Mind in society: The development of higher psychological processes*. Cambridge, MA: Harvard University Press.

Warden, C. A., Chen, J. F., & Caskey, D. (2005). Cultural values and communication online: Chinese and Southeast Asian students in a Taiwan international MBA class. *Business Communication Quarterly, 68*(2), 222–232.

Zhu, Y., Gareis, E., Bazzoni, J. O., & Rolland, D. (2005). A collaborative online project between New Zealand and New York. *Business Communication Quarterly, 68*(1), 81–96.

Contributors

Carol M. Barnum, PhD, is director of graduate programs in information design and communication at Southern Polytechnic State University, Marietta (Atlanta), Georgia, where she teaches a variety of courses, including international technical communication. Barnum is a Fulbright senior specialist with expertise in business and technical communication in Asia, a fellow of STC, an STC Gould Award winner for excellence in teaching technical communication, and the author of six books and numerous articles on business and technical communication and usability research.

Audrey G. Bennett received a BA in studio art from Dartmouth College and an MFA (a terminal degree) in graphic design from Yale University's School of Art. She is currently an associate professor in the Department of Language, Literature, and Communication at Rensselaer, where she teaches and conducts research on graphics. Her research interests include design theory and research for cross-cultural communication and social change. She is the editor of *Design Studies: Theory and Research in Graphic Design*, published by Princeton Architectural Press, which chronicles the historical and contemporary efforts of designers to broaden the scope of the profession of graphic design to include research.

Paul Blewchamp has taught English in the United Kingdom and Taiwan. He is currently a lecturer in English at Shih Chien University, Taiwan, where he specializes in the teaching of spoken discourse.

Tsui-ping Chen is a PhD student in the English Department of the National Kaohsiung Normal University and a lecturer in the Applied English Department of Kun Shan University, Taiwan. Her research is focused on EFL/ESL reading and writing.

Boyd Davis, a linguist, is Cone Professor at the University of North Carolina–Charlotte and was a visiting professor at the National Kaohsiung Normal University, Taiwan. She develops digital corpora of first- and second-language speakers of English and of multilingual elders with and without cognitive impairment from Alzheimer's disease; she uses the corpora to analyze language and develop educational materials.

Daniel Ding has a PhD in English with emphasis in rhetoric and technical communication. He is a professor of composition and technical communication at Ferris State University in Big Rapids, Michigan, where he teaches technical communication, scientific writing, and advanced composition at Ferris State University, Big Rapids, Michigan. His research interests include the history of technical communication, international technical communication, the history of scientific writing, and multiculturalism in composition and technical communication. He publishes in these areas. In addition, he often gives lectures on technical communication and composition in foreign universities. Daniel Ding would like to express his appreciation to the editors of this book, Kirk St.Amant and Filipp Sapienza, and Baywood's anonymous editors for their generous comments and suggestions on the drafts of this chapter.

Ron Eglash received a BS in cybernetics, an MS in systems engineering, and a PhD in the history of consciousness, all from the University of California. A Fulbright postdoctoral fellowship enabled him to conduct field research on African ethnomathematics; the results of his research were published by Rutgers University Press in 1999 as *African Fractals: Modern Computing and Indigenous Design.* He is now an associate professor of science and technology studies at Rensselaer. He teaches courses including a studio class on the design of educational technologies and graduate seminars on the relations between science and society. His research interests include ethnomathematics, complexity theory, social studies of science, race, and racism, and cybernetics.

R. Peter Hunsinger is completing his doctoral work at Iowa State University. He plans to finish by May 2010. His research interests include cultural theory, globalization, and the implications of global communication technologies for technical and professional communication practice.

Sipai Klein is a visiting Assistant Professor at University of Colorado–Colorado Springs. He received his doctoral degree in Rhetoric and Professional Communication from New Mexico State University, an MA in English Literature from the City College of New York, and a BA in Physics from Yeshiva University. His research focuses on new media, multimodality, and digital rhetoric.

Mukkai S. Krishnamoorthy received a BE degree with honors from Madras University and an MTech degree in electrical engineering and a PhD from the Indian Institute of Technology, Kanpur. His research interests are in the design and analysis of combinatorial and algebraic algorithms, visualization algorithms, problem-solving paradigms, and programming environments. He is currently an associate professor of computer science at Rensselaer.

Sharon Trujillo Lalla is currently an adjunct faculty member in education and the instructional technology manager supporting the learning management system at New Mexico State University. She received an EdD in educational technology from Pepperdine University. In 1987, she received her MA degree, specializing in technical communication, from New Mexico State University. She has been an educator, technical communicator, and editor in both industry and academia throughout her professional career. She has taught traditional, blended, and totally online courses in higher education for over 15 years. Her published articles and presentations cover topics such as local house styles, the TEACH act, constructivism online, online communications, and successful training programs.

Clinton R. Lanier is an assistant professor of technical communication at the New Mexico Institute of Technology. Prior to pursuing his doctorate, he worked as a technical writer for the software industry and as a technical editor for the U.S. Army Research Laboratory at White Sands Missile Range, New Mexico. He received a PhD in rhetoric and professional communication from New Mexico State University in 2006. His research interests are in software documentation, computer science education, usability, and international professional communication.

Matthew McCool teaches writing at Southern Polytechnic State University in Atlanta, Georgia.

Hui-Fang Peng is currently a lecturer in English at Fortune Institute of Technology and a PhD student at National Kaohsiung Normal University in Taiwan. She specializes in the teaching of reading and listening

Martine Courant Rife has an active license to practice law in Michigan. She is a PhD candidate in rhetoric and writing at Michigan State University and is a member of the writing faculty at Lansing Community College. Her research is on intellectual property law, technical writing, and rhetoric in digital environments. Martine's work has been published in *Kairos, Pedagogy, Teaching English in the Two Year College, Technical Communication, Computers and Composition,* the *Technical Communication Quarterly,* and the *Journal of Business and Technical Communication.* She has book chapters in several edited collections. She is the 2007 winner of the Society for Technical Communication's Frank R. Smith Outstanding Journal Article Award.

Kirk St.Amant is an associate professor of technical and professional communication at East Carolina University, where he teaches courses in professional communication in international contexts. His research interests include international online communication, international virtual workplaces, and international outsourcing practices.

Filipp Sapienza is a usability and content management consultant based in Denver, Colorado. He holds a doctorate from Rensselaer Polytechnic Institute and publishes research on usability, information design, and cross-cultural Web design.

Judith B. Strother, an associate fellow of IEEE, is chair of the graduate programs in communication and professor of applied linguistics at Florida Institute of Technology, Melbourne, Florida. Her PhD in technology and linguistics is from the Technische Universiteit Eindhoven in the Netherlands, where she is a part-time visiting professor. She also has an MA in applied linguistics and an MBA. Her research interests include intercultural business communication, Web-based language training, service leadership, and new technologies for marketing communication. She chairs a development team that provides Web-based aviation English courses.

Index

Selected titles from

Baywood's Technical Communications Series

Series Editor, *Charles H. Sides*

ASSESSMENT IN TECHNICAL AND PROFESSIONAL COMUNICATION
Edited by *Margaret Hundleby and Jo Allen*
— Winner of the CPTSC Award for Excellence —
in Program Assessment

CONTENT MANAGEMENT
Bridging the Gap Between Theory and Practice
Edited by *George Pullman and Baotong Gu*

CONNECTING PEOPLE WITH TECHNOLOGY
Issues in Professional Communication
Edited by *George F. Hayhoe and Helen M. Grady*

OUTSOURCING TECHNICAL COMMUNICATION
Issues, Policies and Practices
Edited by *Barry L. Thatcher and Carlos Evia*

MOTIVES FOR METAPHOR IN SCIENTIFIC AND TECHNICAL COMMUNICATION
Timothy D. Giles

TOGETHER WITH TECHNOLOGY
Writing Review, Enculturation and Technological Mediation
Jason Swarts
— Winner of the 2009 NCTE Award for Best Book —
in Technical or Scientific Communication

RESOURCES IN TECHNICAL COMMUNICATION
Outcomes and Approaches
Edited by *Cynthia L. Selfe*
— Winner of the 2007 STC Award of Excellence —
NY Metro Chapter

COMMUNICATIVE PRACTICES IN WORKPLACES AND THE PROFESSIONS
Cultural Perspectives on the Regulation of Discourse and Organizations
Edited by *Mark Zachry and Charlotte Thralls*
— Winner of the 2008 NCTE Award for Best Collection of Essays —
in Technical or Scientific Communication

ONLINE EDUCATION
Global Questions, Local Answers
Edited by *Kelli Cargile Cook and Keith Grant-Davie*
— Winner of the 2006 NCTE Award for Best Collection of Essays —
in Technical or Scientific Communication

For Product Safety Concerns and Information please contact our EU
representative GPSR@taylorandfrancis.com
Taylor & Francis Verlag GmbH, Kaufingerstraße 24, 80331 München, Germany

www.ingramcontent.com/pod-product-compliance
Ingram Content Group UK Ltd.
Pitfield, Milton Keynes, MK11 3LW, UK
UKHW022001180926
475701UK00003B/129

* 9 7 8 0 4 1 5 4 0 3 1 8 4 *